Lecture Notes in Mathematics

Edited by A. Dold and B. Eckmann

1174

Categories in Continuum Physics

Lectures given at a Workshop
held at SUNY, Buffalo 1982

Edited by F. W. Lawvere and S. H. Schanuel

Springer-Verlag
Berlin Heidelberg New York Tokyo

Editors

F. William Lawvere
Stephen H. Schanuel
State University of New York at Buffalo
Mathematics Department, 106 Diefendorf Hall
Buffalo, N.Y. 14214, USA

Mathematics Subject Classification (1980): 18-xx, 53-xx, 58-xx, 70-xx, 76-xx, 80-xx

ISBN 978-3-540-16096-0 Springer-Verlag Berlin Heidelberg New York Tokyo

Library of Congress Cataloging-in-Publication Data. Categories in continuum physics. (Lecture notes in mathematics; 1174) Bibliography: p. 1. Field theory (Physics) – Addresses, essays, lectures. 2. Continuum mechanics – Addresses, essays, lectures. I. Lawvere, F. W. II. Schanuel, S. H. (Stephen Hoel), 1933-. III. Series: Lecture notes in mathematics (Springer-Verlag); 1174. QA3.L28 no. 1174 [QC174.46] 510 s [530] 86-1806

2146/3140-543210

Preface

The success of the May 1982 Workshop which resulted in this volume was due in large measure to the active and friendly participation of all who attended, to the valued assistance of Rosemary Marciniak and Judy Bittner, and to the financial assistance granted by SUNY Buffalo. We thank all these, and especially thank Fatima Lawvere, whose dedicated efforts brought about the atmosphere which all enjoyed.

For the preparation of this volume, the editors give thanks to Gail Berti for her excellent typing, and to the National Science Foundation for financial support.

F. William Lawvere
Stephen H. Schanuel

C O N T E N T S

INTRODUCTION

The articles collected in this volume reflect talks given at a workshop on Category Theory and the Foundations of Continuum Thermomechanics which was held at S U N Y Buffalo in May 1982. The workshop permitted the beginning of more extensive exchange of ideas between groups of researchers which had previously had very little contact. W. Noll and W. Williams discuss here the foundations of the theory of material bodies, in particular the current status of Cauchy's stress theorem, while B. Coleman, M. Feinberg, R. Lavine, and D. Owen describe general contexts in which the existence of temperature and entropy have been established for rapidly deforming, unequally heated bodies. In apparently different directions, K.T. Chen and A. Frölicher discuss simplification of the foundation of infinite-dimensional differential geometry, as do A. Kock and G. Reyes with special attention to the axiomatics of that subject as a whole and to the variety of useful models of the resulting theory. The geometric theory is necessarily categorical, and I want to indicate below some of the advantages which may result if the abstract structures arising in thermomechanics are also explicitly recognized as categories. First I will indicate some of the reasons why a flexible geometric theory is demanded by continuum physics.

The mathematical background for theories of geometry, analysis, and continuum physics is usually considered to be the category of topological spaces or the category of Banach manifolds, with of course an infinite gradation of smoothness conditions needed (apparently) for various technical theorems. However, an essential construction in continuum physics is that of "function space", and the lack of well-behaved function spaces in those categories obscures the simplicity of geometrical or physically-motivated constructions and axioms. Yet, two centuries ago, many problems in the calculus of variations were correctly solved by mathematicians who, rather than defining a notion of "open subset" for their function spaces, took the notion of "path" as basic. Recognizing the great importance of contravariant concepts such as open set (or real function) does not commit us to take these as the <u>defining</u> structure of a notion of space-in-general; they can be <u>derived</u> concepts in a theory where the covariant concept of geometric <u>figures</u> of some basic types, such as path, tangent vector, etc. are taken as primitive; theories of the latter kind can easily be constructed in which the unambiguous function space construction with good properties exists. It is with the construction of such categories that the articles of K. T. Chen, G. Reyes, and A. Frölicher are concerned. The "figures" used by Chen are called by him "plots", extending the less functorial charts of the usual atlases. Another significant aspect pointed out by these authors is that working in a category of smooth morphisms does not at all restrict one to considering only smooth subobjects in the usual manifold sense.

More precisely, Chen and Frölicher study two specific categories in which profound basic theorems about the smooth real line are used to construct notions of space which encompass many necessary examples that are not Banach manifolds, and are far simpler to describe from first principles than is the latter notion.

Kock and Reyes emphasize the axiomatic description of such categories as a whole. Axiomatizing a category as a whole promises to be part of the simplest approach to certain calculations. One exploits the discovery of Grothendieck that once the covariant "figure" attitude toward spatial objects is adopted, not only function spaces but also several other mysterious notions become easily manageable, such as infinitesimal paths, the spatial structure of the "set" of all linear subspaces of a given linear space, etc. Moreover, a "category of all spaces" can be construed as a "gros topos" which implies that fibered products and quotient constructions have exactness properties similar to those in the naive category of abstract sets but lacking in the usual categories of topological spaces or manifolds. The axiomatics at the category level is also valuable because there are many related categories which immediately come up. For example, if $\mathcal{X}$ is a gros topos of spaces and G is a group in $\mathcal{X}$ while S is an object of $\mathcal{X}$ then the categories $\mathcal{X}/S$ of S-parameterized families of spaces, $\mathcal{X}^G$ of actions of G on spaces in $\mathcal{X}$, and $\mathcal{X}^G/S$ (of central interest in bifurcation theory) are all categories which satisfy the same axioms as $\mathcal{X}$, as does (a reasonable determination of) the category of all objects of $\mathcal{X}$ equipped with affine connection.

Let me be more explicit about the role of the cartesian-closed property of a category (a <u>topos</u> is a cartesian-closed category in which moreover the notion of subobject is representable by a "truth-value" object). Let E denote ordinary physical space, T a space which represents the notion of time, and B a space which represents a particular body. Then a particular motion of B may be represented as a map

$$B \times T \longrightarrow E$$

which is the correct way if we want to compute by composition how particles of the body at various times experience the values of some field defined on space. However, it is also necessary to construe the same motion as a map

$$T \longrightarrow E^B$$

where the space E^B of (possibly singular) placements of the body is itself independent of T or a particular motion, if we want to compute by composition the temporal variation of quantities like the center of mass $E^B \longrightarrow E$ of B. Still a third version

$$B \longrightarrow E^T$$

of the same motion, where the space E^T of paths in space exists independently of B, is a necessary step if we want to compute by composition the velocity field on B induced by the motion. The possibility of passing freely among these three versions of the "same" map is obviously more fundamental for phrasing general axioms and concepts of continuum physics than is the precise determination of the concept of spaces-in-general (of which E,T,B are to be examples), yet these transformations are <u>not</u> possible for the commonest such determinations (for example Banach manifold). The general possibility of such transformations within a given category is called cartesian closure; a category with finite cartesian products (including an empty product 1) is cartesian closed if for any two objects A,Y there is another Y^A such that for any object X there is a natural bijection

$$\frac{X \longrightarrow Y^A}{A \times X \longrightarrow Y}$$

of maps (= morphisms in the given category). In particular any functional $Y^A \longrightarrow Z$ when composed with any $I \longrightarrow Y^A$ gives a map $I \longrightarrow Z$. A lemma proved by Grothendieck and by Yoneda (and in special cases by Cayley and Dedekind) says in effect that an object in a category is entirely determined by all morphisms into it from all possible objects. But in many categories there is a small set of objects such that morphisms from them alone into an arbitrary object determine the latter. Such special objects I might be called generic figures, and morphisms $I \xrightarrow{x} X$ particular figures of type I in X. If the I's are adequate in the sense [Isbell] just alluded to, a morphism $X \xrightarrow{f} Y$ is determined by the abstract mapping $x \longmapsto fx$ of figures in X into figures in Y; and more importantly, such an abstract mapping is "smooth" (i.e. comes from an actual morphism $X \longrightarrow Y$ in the category) if only it is natural, i.e. satisfies the property $f(xa) = (fx)a$ for all $I' \xrightarrow{a} I$ between figures only (one might say that the mapping preserves generalized incidence relations). Such reasoning is sufficient to rigorously support the calculations of the Calculus of Variations, taking intervals as the adequate generic figures: to discuss the smoothness of a purported functional

$$Y^A \xrightarrow{J} Z$$

it suffices to check its compositions with all $I \xrightarrow{\bar{v}} Y^A$, but the latter are equivalent to $A \times I \xrightarrow{v} Y$ which are of lower type.**) Since the v are the origin of the term "variation" in "Calculus of Variations" we may say that the combination of the notion of cartesian closed category with that of generic figure to yield a determination of "space-in-general" is a natural development of those 18th century ideas. For convincing substantiation of these apparently simple-minded remarks, see the work of Chen and Frölicher.

In the articles of Kock and Reyes an important additional feature is the consideration of further generic figure types such as D, the tangent vector. This object is explicitly definable in terms of the smooth line R, as the subspace of R consisting of all $t \in R$ for which $t^2 = 0$. This object D can be non-trivial without changing the morphisms $R^n \longrightarrow R^m$, which remain the usual smooth maps (or the usual analytic or algebraic maps in other models of the axioms). In fact, D can be big enough that

$$R^D \xleftarrow{\sim} R \times R$$

in the sense that every smooth function defined near 0 restricts to equal a <u>unique</u> affine-linear function on D. Identifying the function space X^D with the tangent-bundle of an arbitrary space X in the category, a vector field $X \longrightarrow X^D$ becomes equivalent, via the fundamental transformation, to an action

$$D \times X \xrightarrow[\xi]{} X \qquad \xi(0,x) = x$$

of D on X. An obvious way for such to arise is to restrict some flow (action of the additive group of R)

$$R \times X \longrightarrow X$$

**) The smoothness of morphisms whose domain is a product $A \times I$ can be analyzed by testing against arbitrary $I \longrightarrow A \times I$ as is further explained in Frölicher's article.

from R to D , such restriction defining an instance of differentiation. This gives rise to a pair of adjoint functors

$$\text{Vector-fields} = \mathcal{X}^{D} \overset{(\ )^{\bullet}}{\underset{\longrightarrow}{\longleftarrow}} \mathcal{X}^{R} = \text{Flows}$$

where the right adjoint to differentiation takes X,ξ to

$$\mathrm{Hom}_D(R,X)$$

the space of solution curves. The adjunction map of flows, evaluation at 0 , is an isomorphism when the vector field given on X satisfies the existence and uniqueness properties as an ODE. It is worth pointing out that the definition of morphism ψ in the category $\mathcal{X}^D$ of vector fields, namely commutativity of

$$\begin{array}{ccc} D \times X & \xrightarrow{D \times \psi} & D \times Y \\ \Big\downarrow \xi & & \Big\downarrow \eta \\ X & \xrightarrow[\psi]{} & Y \end{array}$$

is equivalent by the fundamental transformation to commutativity of

$$\begin{array}{ccc} X^D & \xrightarrow{\psi^D} & Y^D \\ \bar{\xi} \Big\downarrow & & \Big\downarrow \bar{\eta} \\ X & \xrightarrow[\psi]{} & Y \end{array}$$

which makes sense even in the usual category of manifolds where the tangent bundles are not function spaces in the way they are here. To deal with non-autonomous flows, the monoid R can be replaced by a suitable small category whose objects are instants of time.

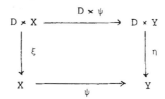

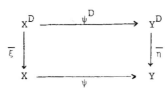

Some fundamental concepts of continuum physics can be formulated in such categories before determination of which category is most appropriate for special calculations, or even in some cases before distinguishing between vector fields $\mathcal{X}^D$ and flows $\mathcal{X}^R$. For example, Muncaster's clarification [Muncaster] of the general problem of deriving coarse theories from fine theories is (backed up by his study of several key cases) in essence the following (with M = D or M = R): Let X be an object of fine states equipped with a dynamics, i.e. $X \in \mathcal{X}^M$ with action ξ , and let $Y \in \mathcal{X}$ be an object of coarse states with an $\mathcal{X}$-map $X \xrightarrow{\pi} Y$ (often an averaging process of some sort). It is desired to find dynamical actions η on Y which are somehow compatible with ξ and π , but experience shows that this should <u>not</u> mean that π becomes an $\mathcal{X}^M$ morphism for the choice of η ; π may preserve balance laws but not constitutive relations, both ingredients being involved in specifying η. Call an M-action η a non-linear "eigenvalue" of ξ (with multiplicity Y), if we can find an injective M-morphism G (in the direction <u>opposite</u> to π)

$$\begin{array}{ccc} M \times Y & \xrightarrow{1_M \times G} & M \times X \\ \eta \downarrow & & \downarrow \xi \\ Y & \xrightarrow[G]{} & X \end{array} \qquad G\eta = \xi(1_M \times G)$$

where such G could be called a non-linear "eigenvector". A <u>gross determiner</u> is such a non-linear eigenvector which (interprets coarse states as special fine states as above and) satisfies the constraint $\pi \circ G = 1_Y$ (which of course forces G to be injective). It is to be expected that there may be many such eigenpairs η, G for a given fine theory X, ξ and given "averaging" π to coarse states Y; but the gross determiner G uniquely determines the coarse theory η (as with the usual linear eigenvalues).

The adjointness above (between differentiation of flows and solution curves) actually holds for any change of operator domain (like $D \subset R$), for example for the change $1 \xrightarrow{0} \mathbb{N}$, where 1 is the trivial domain and $\mathbb{N}$ is the additive monoid of natural numbers. An $\mathbb{N}$-action is equivalent just to a single endomorphism thought of as the change of state in one time unit. The adjunction in this case

$$\mathcal{X} \underset{(\ )^{\mathbb{N}}}{\overset{\text{forget action}}{\rightleftarrows}} \mathcal{X}^{\mathbb{N}}$$

is simply this: if Y is any object, then the object $Y^{\mathbb{N}}$ of sequences in Y has a standard action usually called "shift". If X is an object equipped with <u>any</u> endomorphism ξ, then an <u>equivariant</u> map

$$X \xrightarrow{\bar{\psi}} Y^{\mathbb{N}}$$

is entirely determined by an ordinary map

$$X \xrightarrow{\psi} Y$$

(namely $\psi(x) = \bar{\psi}(x)_0$) through the formula

$$\bar{\psi}(x)_n = \psi(\xi^n x) \qquad \text{all } n,$$

and for arbitrary ψ, the $\bar{\psi}$ so defined is in fact equivariant. This is sometimes referred to as "symbolic dynamics", Y being considered as blocks into which X is divided by ψ, and the $\bar{\psi}$ thus assigning to a state x the sequence of blocks through which the dynamics ξ

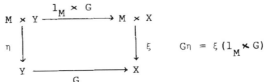

takes x . In case Y is finite, $Y^{\mathbb{N}}$ is a Cantor Space in $\mathcal{X}$. The basic concept (which can be further ramified) of the currently popular "chaos" is that of a morphism ψ for which the induced $\bar{\psi}$ to the right adjoint is surjective (i.e. ψ observes so little of the states x that any given sequence of blocks can occur for some choice of x).

The important distinction between intensive and extensive quantities can also be exemplified in any category $\mathcal{X}$ of the kind under consideration. While these terms, of philosophical origin, are costumarily employed only in thermodynamics, (contrasting temperature, pressure, and density with energy, volume, and mass), they are actually applicable throughout continuum physics and indeed in mathematics generally. While their importance is most evident when interpreted relative to a given body B , it is useful to consider intensive and extensive quantities relative to any space X . The existence of this philosophical terminology is moreover fortunate because terms like "Radon measure" , "Schwartz distribution" , "singular homology class", etc. prejudice the issue in that they are but realizations of the general notion of extensive quantity resulting from various particular determinations of the spatial category $\mathcal{X}$. If we suppose given a ring object R in $\mathcal{X}$ then a basic notion of intensive quantity relative to an object X is that of a morphism $X \longrightarrow R$. Thus R^X is the space of intensive quantities on X . Two distinguishing features of intensive quantity are contravariance: there are

$$R^X \longleftarrow R^Y$$

induced for any $X \longrightarrow Y$, and multiplicativity: R^X is again a ring object (not just an R-linear space) and the foregoing induced morphisms $R^X \longleftarrow R^Y$ are ring homomorphisms (not merely R-linear). By contrast extensive quantity $M(X)$ should be covariant:

$$M(X) \xrightarrow[\psi!]{} M(Y)$$

exists induced by any $X \xrightarrow{\psi} Y$ and is linear but not multiplicative. (A ring structure on $M(X)$ can often be defined via convolution if X is a group or monoid, but not if X is just a space). However, the linearity of M is stronger than just R-linearity in the sense that $M(X)$ is actually a module over the ring R^X and the induced maps $M(X) \longrightarrow M(Y)$ are linear with respect to that, i.e. for $X \xrightarrow{\psi} Y$

$$\psi!(g\,\psi\cdot m) = g\cdot\psi!(m)$$

for any $g \in R^Y$, $m \in M(X)$. Moreover, there is a pairing

$$R^X \times M(X) \xrightarrow{\int_X} R$$

between intensive and extensive quantity which satisfies naturality equations when X is varied, and

$$\int_X f\,d(g\cdot m) = \int_X (fg)\,dm \qquad \text{for all } f \ .$$

All these listed properties of M follow easily if we simply define

$$M(X) = \mathrm{Lin}_R(R^X, R)$$

the space of $\mathcal{X}$-smooth linear functionals, where in general $\mathrm{Lin}_R(V,W)$ is the subobject of W^V defined by the linearity equations, for any R-linear spaces V,W in $\mathcal{X}$. Then $\int$ is simply evaluation and the module structure is <u>defined</u> by the above fg equation. This definition of M can be proved (in all the categories considered by Chen, Frölicher, Kock and Reyes) to coincide with usual distributions with compact support, even though smoothness in those categories is defined covariantly in terms of generic figures rather than contravariantly in terms of open sets or seminorms. In other toposes it would agree rather with topological or abstract measure theory. The module structure expresses the important concept of density: if m,v are two quantities extensive with respect to the same space, then

$$\frac{dm}{dv} = \rho$$

simply means that ρ is an intensive quantity such that $m = \rho \cdot v$ for the module structure. Leaving to particular determination of the category the question of precisely which pairs m,v admit such a ρ, we may observe that its "uniqueness" (in a natural sense) is tautological: if $\rho_1 \cdot v = \rho_2 \cdot v$, then $(\rho_1 - \rho_2)\cdot v = 0$ i.e. $\rho_1 \equiv \rho_2$ modulo v. It will be seen that the covariant functoriality of extensive quantity is an essential background feature permitting Feinberg and Lavine's passage from states to thermodynamic states and Schanuel's passage from the size of a potato to its polynomial measure on space.

The categorical striving for unity and simplicity may also lend clarification to problems of continuum physics in another way. Beyond the background questions where the nature of categories like spaces-in-general, flows-in-general, etc. is studied, there is also the observation that, like many other branches of mathematics, thermomechanics deals with structures which themselves may sometimes usefully be seen as ("small") categories. The article by D. Owen is a case in point.

The general theory of thermodynamical systems developed by him and Coleman has features equally applicable both globally to a body as a whole and infinitesimally to body elements, and so there arises the crucial problem of "integration", i.e. of understanding how the body can glue the infinitesimal thermodynamical systems to obtain the global one. Owen proposes to approach this problem through the notion of sheaf, that is, by studying a certain kind of functor from a category of parts of the body into a category of abstract (or topological,or bornological) thermodynamical state-and-process systems. I am convinced that this line of thought will become important. I will comment separately on both the category of parts and the category of systems, with special reference to the utility of considering the systems themselves as categories.

The theory of parts of a body, discussed as a necessary preliminary in the articles of Noll and Williams, naturally concentrates on the subbodies (which might with luck form a Boolean algebra) but also must take account of boundaries (which are not sub-bodies). A convenient algebraic structure which includes both these features is that of a cartesian-closed partially-ordered set in which " $\longrightarrow$ " is thought of as " $\supseteq$ " and hence "cartesian product" becomes $\cup$ while "exponentiation" becomes a binary operation akin to subtraction, which is characterized by

$$A \supseteq C\setminus B \quad \text{iff} \quad A \cup B \supseteq C$$

The resulting algebraic system may alternatively be described as a lattice admitting such a subtraction; the subtraction is unique if it exists and its existence implies distributivity of the lattice operations. The system of all closed subsets of a given toplogical space is a typical example to keep in mind; the subtraction operation in that case is forced to be the closure of the set-theoretic difference. However, examples not of that form arise in various parts of mathematics, and it might reasonably be hoped that models of bodies which involve entities more (or less!) sophisticated than closed sets would still admit the structure here discussed. If 1 denotes the whole body, then we can define

$$\sim A = 1\setminus A$$

as a special case of the subtraction; $\sim A$ is thus characterized as the smallest object in the system for which $\sim A \cup A = 1$. One always has $\sim(A \cap B) = \sim A \cup \sim B$, but it can happen that $\sim(A \cup B) \neq \sim A \cap \sim B$; however $\sim\sim(A \cup B) = \sim\sim A \cup \sim\sim B$ always holds. Then

$$\sim\sim A \subseteq A$$

is the <u>regular</u> core of A, so that, with Noll and Williams (and Tarski), we may consider a part A of 1 to be a sub-body, or simply a body, if and only if $\sim\sim A = A$. But in such a lattice we may also define

$$\partial A = A \cap \sim A$$

and consider this as the boundary of A in the sense of the system considered. That the notion of boundary is just that of "logical contradiction" (within the realm of closed sets) follows at once from the intuitive notion of motion: indeed, since the unit interval is connected, any continuous path which is in A at time 0 and in $\sim A$ at time 1 must at some intermediate time be in both A and $\sim A$, i.e. must pass through the boundary of A. Independently of the motivating example of closed parts, a great many useful identities can be proved in general for any lattice which satisfies our axiom of subtraction. For example

$$A = \sim\sim A \cup \partial A$$

for all A, and (as pointed out by R. Flagg)

$$\partial(A \cap B) = ((\partial A) \cap B) \cup (A \cap \partial B)$$

for all A,B. The latter "Leibniz formula" is remarkable in that, although it is pictorially obvious

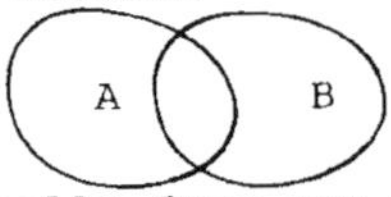

and easily proved algebraically from our axioms (or in particular from the definition of closed set), and although similar formulas are well-known for more sophisticated objects such as currents, etc. the fact that it is true for ordinary boundaries of ordinary closed sets seems to have escaped the authors of textbooks on general topology; indeed the only source we could find for it (for which I thank G. Rousseau) is a little-known article by M. Zarycki (Fund. Math. 1927). Also evident from the above picture, and valid in any system of the kind under discussion, is

$$\partial(A \cup B) \cup \partial(A \cap B) = \partial A \cup \partial B$$

so that in particular

$$A \cup B = 1 \Longrightarrow \partial(A \cap B) = \partial A \cup \partial B$$
$$(\partial A) \cap B = 0 = A \cap \partial B \Longrightarrow \partial(A \cup B) = \partial A \cup \partial B$$

The special parts which are boundaries form an ideal (of "nowhere dense" parts), whose elements can also be characterized by many alternative equations, for example

$$\partial A = A, \text{ or } \sim A = 1 \text{ , or } \sim\sim A = 0, \text{ or } A \cup Y \subseteq \sim A \cup Y$$

for all Y. Note that $\partial \sim A = \partial \sim\sim A$ is in general smaller than ∂A; equality of all three can hold only if A is a body.

A crucial relation between sub-bodies is that they be separate, or equivalently that they be (at most) in contact. For $\sim\sim A = A$, $\sim\sim B = B$, the following possible definitions of this relation are equivalent:

$$\sim B \supseteq A$$
$$\sim A \supseteq B$$
$$\sim\sim (A \cap B) = 0$$
$$A \cap B = \partial A \cap \partial B$$

By an interaction is meant a function H (in general vector valued) defined for separate pairs of bodies and such that $H(A,-)$ is additive on separate pairs and likewise $H(-,B)$. The importance of the material point of view having been recognized, it would be desirable to analyze, insofar as possible without reference to the momentary embeddings of the body in space, properties of interactions such as the property of being a surface interaction. By a surface interaction I here mean an interaction H such that whenever A_1, A_2 are bodies separate from the body B

$$A_1 \cap B = A_2 \cap B \Longrightarrow H(A_1, B) = H(A_2, B)$$

and similarly in the other variable (the last being automatic in case $H(A,B) = -H(B,A)$ holds.) This turns out to be equivalent to the vanishing of H on certain pairs. Namely, call a pair D,B very separate (or in at most slight contact) if

$$\sim (D \cup B) \supseteq D \cap B$$

or equivalently both

$$\partial (D \cup B) = \partial D \cup \partial B \quad \text{and} \quad \partial (D \cap B) = \partial D \cap \partial B.$$

Then an interaction H is a surface interaction iff $H(D,B) = 0$ whenever D,B are very separate. The crucial constructions for the proof of the foregoing statement are the definition of two "differences"

$$D_i = \sim\sim ((A_1 \cup A_2) \cap (\sim A_i))$$

which will be very separate from B whenever A_1, A_2 are bodies having equal contact with B , and conversely, the definition of a special pair

$$A_1 = \sim(D \cup B)$$
$$A_2 = A_1 \cup D$$

which will have equal contact with B whenever D is a body very separate from B . It seems that still more results could be obtained within the intrinsic "body" point of view, without involving absolute continuity with respect to a surface measure which depends on the instantaneous embeddings of the body in space.

A system of states and processes can also be usefully construed as a small category $\underline{X}$, in which domain and codomain are simply the beginning and end states of a given process and composition is simply the operation of following one process by another. A category of this sort is often equipped with a "duration" functor to the additive monoid of nonnegative time translations; the functoriality is just the equation

$$\mathrm{dur}(\beta\alpha) = \mathrm{dur}(\beta) + \mathrm{dur}(\alpha)$$

for composable processes, and

$$\mathrm{dur}(1_X) = 0$$

for the identity process of any state x . In examples the functor dur satisfies a further axiom of "unique lifting of factorizations" as follows

$$\mathrm{dur}(\gamma) = s + t \Longrightarrow \exists!\ \alpha, \beta\ [\beta\alpha = \gamma,\ \mathrm{dur}(\beta) = s,\ \mathrm{dur}(\alpha) = t]$$

When the latter condition holds, the state with which α ends could be considered as the state " $\gamma(t)$ " through which γ passes at the intermediate time t . Thus each morphism (process) γ in such a category $\underline{X}$ determines a path through the objects (states), and indeed a large class of examples can be constructed by starting with an object X in a topological category $\mathcal{X}$, defining the objects of $\underline{X}$ to be the points of X and defining the morphisms of $\underline{X}$ to be arbitrary continuous paths in X whose domains are intervals [0,a] . Again the presumption that $\mathcal{X}$ is a topos assists in constructing the totality of such paths as a single object (if needed):

$$\begin{array}{ccc} P(X) & \longrightarrow & \tilde{X}^R \\ \downarrow & \text{pullback} & \downarrow \mathrm{dom}^R \\ R^+ & \longrightarrow & \tilde{1}^R \end{array}$$

where $\tilde{X}$ is the partial-map classifier which exists in any topos, $\tilde{1}^R$ is the power "set" of R , and the lower map is the one which assigns $[0,a]$ to any a . The fact that $PX \rightrightarrows X$ is a category follows from the pushout property of intervals in the <u>topological</u> topos:

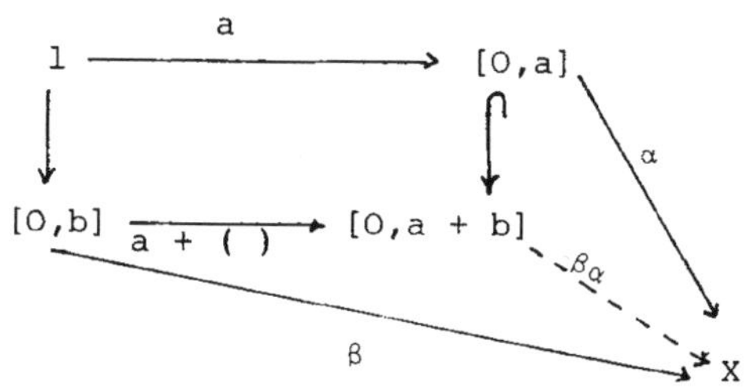

It is essentially the failure of this pushout property (i.e. of the failure of paths to be closed with respect to the indicated composition) in a smooth (as opposed to continuous) topos which forced historically the introduction of <u>piecewise</u> smooth paths; the possibility of collecting even these into one "smooth" object is one of the important ingredients in the work of Chen. A general explanation for this possibility is the following: the <u>subdivided</u> smooth paths can in any case be collected into a "simplicial" object X^* in $\mathcal{X}$ and the inclusion $\mathrm{Cat}(\mathcal{X}) \hookrightarrow \mathcal{X}^{\Delta^{op}}$ has a left adjoint (for any reasonable $\mathcal{X}$), which "completes" a simplicial object to a category. Physically more typical examples of such categories $\underline{X}$ are obtained as non-full <u>sub</u>categories wherein one restricts attention to "admissible" processes obeying some constitutive relation. More precisely, there is often a functor

$$\begin{array}{c} \underline{X} \\ \downarrow{\scriptstyle \pi} \\ \underline{C} \end{array}$$

from states to "configurations" which satisfies determinism in the form of the categorical <u>fibration</u> condition: Given any object (state) x in $\underline{X}$ and any morphism (deformation process) γ in $\underline{C}$ such that γ starts at $\pi(x)$, there is a unique morphism (state process) $\overline{\gamma}$ in $\underline{X}$ for which $\pi(\overline{\gamma}) = \gamma$ and $\overline{\gamma}$ starts at x . Under this condition clearly $\underline{X}$ is not usually a full category of paths even if $\underline{C}$ may be. Note that following such a fibration π with a duration functor for $\underline{C}$ will provide a duration functor for $\underline{X}$. For a given category $\underline{C}$ equipped with a duration there is a canonical example [Noll] of such $\underline{X}$ (of importance in the theory of materials with fading memory) in which the objects of $\underline{X}$ are taken to be <u>histories</u> in $\underline{C}$. Here a history is any functor $[0,\infty)^{op} \xrightarrow{x} \underline{C}$ (from the reversed <u>ordered</u>

set of nonnegative time translations) for which $\mathrm{dur} \circ x = \Delta$, where Δ is the "difference" functor from the ordered set of time translations to the monoid of time translations. Morphisms between histories are just arbitrary natural transformations of such functors, and a functor π can be defined as the restriction (along the indicated full inclusion) of the "evaluation at 0" functor

$$\mathrm{Hist}(\underline{C}, \mathrm{dur}) \hookrightarrow \underline{C}^{[0,\infty)^{op}} \xrightarrow{ev} \underline{C}, \qquad \pi = ev \circ (\text{inclusion})$$

That π satisfies the above fibration condition follows from the "unique lifting of factorizations" property assumed for the duration functor on $\underline{C}$, using the fact that the time-translation-monoid is 1) commutative and 2) cancellative; for then 1) given any $\cdot \xrightarrow{\zeta} \cdot \xrightarrow{\gamma} \cdot$ in $\underline{C}$ there are unique γ', ζ' in $\underline{C}$ such that

$$\begin{array}{ccc} \cdot & \overset{\gamma'}{\dashrightarrow} & \cdot \\ {\scriptstyle \zeta}\downarrow & & \downarrow{\scriptstyle \zeta'} \\ \cdot & \xrightarrow[\gamma]{} & \cdot \end{array} \quad \text{is commutative}$$

and $\mathrm{dur}(\gamma') = \mathrm{dur}(\gamma)$
$\mathrm{dur}(\zeta') = \mathrm{dur}(\zeta)$

and 2) every morphism in $\underline{C}$ is a monomorphism. (Note however that no non-identity morphism in $\underline{C}$ is invertible since 3) $t + s = 0 \Longrightarrow t = 0$ and $s = 0$). There are endomorphisms of non-zero duration which are nonetheless constant as paths - such freezes in a configuration category $\underline{C}$ can have non-trivial dynamical consequences in a state category $\underline{X}$. While when working in a single category it is natural to identify the concept of "cyclic process" with the concept of endomorphism, when comparing two categories by a functor such as $\pi : \underline{X} \longrightarrow \underline{C}$, there is the important phenomenon of hysteresis which must be kept in mind: a morphism $\bar{\gamma}$ in $\underline{X}$ may be such that $\gamma = \pi(\bar{\gamma})$ is cyclic in $\underline{C}$ even though $\bar{\gamma}$ itself is not cyclic in $\underline{X}$. Finally we point out that in many cases it suffices to consider the categories $\underline{C}$ as abstract categories, that is as defined in the topos $\mathcal{S}$ of abstract sets rather than in some higher topological topos $\mathcal{X}$; for a notion of closed set A of configurations (or states) can be derived from the category structure together with the duration functor by requiring that for every γ, $\gamma^{-1}(A)$ is closed in R, and this will often agree with the original topology.

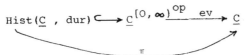

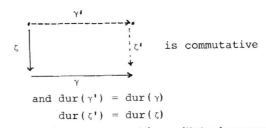

An important way in which the fibration property for a functor π may arise is from a uniqueness and existence theorem for a non-autonomous ordinary differential equation, in which the paths in the lower category $\underline{C}$ represent the time variation of the equation itself, whereas the paths in the upper category $\underline{X}$ are actual solutions. It seems in some contexts needlessly confusing to combine a discussion of this relationship with a discussion of the separate question of the extent to which the paths in $\underline{C}$ themselves are the integrals of their derivatives.

An important role for state categories $\underline{X}$ in thermodynamical theory is to act as domains for quantitative "supplies" (or "actions") $\mathcal{S}$ which are simply functors from $\underline{X}$ into the additive monoid of a linear space or the extended reals. The "additivity" on paths expressed by the functoriality

$$\mathcal{S}(1_x) = 0$$
$$\mathcal{S}(\beta\alpha) = \mathcal{S}(\beta) + \mathcal{S}(\alpha)$$

together with some reasonable continuity condition, suggests that the value of $\mathcal{S}$ at α should be an integral, as indeed it is in many examples; there seems to be a need for representation theorems which would clarify the extent to which a general functor in this context can be expressed by a generalized integral formula. But the functoriality itself, supplemented by assumptions of a qualitative nature, suffices for a great many conceptual results of importance.

When the values of a given supply functor $\mathcal{S}$ are extended reals, particular interest attaches to the possibility of an "entropy" function S of states alone which bounds $\mathcal{S}$ in the sense that

$$\mathcal{S}(\alpha) \leq S(x_2) - S(x_1) \quad \text{whenever} \quad x_1 \xrightarrow{\alpha} x_2 \quad \text{in } \underline{X} .$$

With due attention to the subtleties of addition and subtraction of extended reals (essentially subtraction is <u>adjoint</u> to addition rather than in general inverse to it, so that $[-\infty,\infty]$ becomes a (non-cartesian) <u>closed</u> category, with respect to $\leq$) such an S obviously exists, namely

$$S(x) = \sup\left\{\mathcal{S}(\alpha) \,\middle|\, x_0 \xrightarrow{\alpha} x_1\right\}$$

If $\mathcal{S} < \infty$ and $S(x_0) < \infty$, then $S(x) < \infty$ for all the x for which there exists at least one process $x \dashrightarrow x_0$; the condition that $S(x_0) < \infty$ is equivalent to the Clausius property $S(x_0) = 0$.

Coleman and Owen showed in 1974 that under suitable conditions not only the above naively-defined S but even its upper-semi-continuous regularization $\overline{S}$ bounds $\mathcal{S}$; a formulation of their theorem in the simplified setting described above was recently obtained [Lawvere]. But for each of the diverse special classes of materials to which their theory applies, Coleman and Owen obtained much more, namely that $\overline{S}$ is partially differentiable and that certain equations follow (from the above inequality) which relate the derivatives of $\overline{S}$ to the thermomechanical constitutive relations of the materials. These more precise conclusions have not yet been incorporated into the general theory, but it can be hoped that the present volume contains some of the necessary components for such an advance.

F. William Lawvere

REFERENCES

Isbell, J. R., "Adequate subcategories", Illinois J. Math. 4 541-552 (1960).

Lawvere, F.W., "State Categories, Closed Categories, and the Existence of Semi-Continuous Entropy Functions", IMA Preprint Series # 86, Institute for Mathematics and its Applications, University of Minnesota (1984).

Muncaster, R.G., "Invariant Manifolds in Mechanics I and II, Archive for Rational Mechanics and Analysis, 84, 353-392 (1984).

Noll, W. "A New Mathematical Theory of Simple Materials", Archive for Rational Mechanics and Analysis, 48 1 - 50 (1972). (Also in Noll's Selected Papers, Springer Verlag, 1974).

Truesdell, C. Rational Thermodynamics, 2nd Edition, Springer Verlag (1984).

CONTINUUM MECHANICS AND GEOMETRIC INTEGRATION THEORY

Walter Noll
Department of Mathematics
Carnegie-Mellon University
Pittsburgh, PA 15213

1. Introduction

It is only since the nineteen fifties that some of the basic concepts of continuum mechanics and thermodynamics have been analyzed in the language of contemporary mathematics, i.e., in terms of appropriate mathematical structures. In this paper I will give a brief description of some of these structures. Specifically, I will describe the structure of a material universe (Sect. 2), the concept of an internal interaction (Sect. 3), and the structure of a continuous body (Sect. 4). In Sect. 5 I will describe some of the theory of internal contact interactions, which is at the heart of the mathematical foundations of continuum physics. In order to give this theory a satisfactory form, some serious mathematical difficulties need to be overcome. In Sect. 6 I will summarize these problems and present what I believe the features of a good theory should be. It is fairly evident that the development of such a theory must borrow heavily from the modern mathematical theory of geometric integration.

Most of what I present here is distilled from my Lectures on the Foundations of Continuum Mechanics and Thermodynamics [N], where one can find also most of the relevant references. I have avoided here all consideration of an "external world" and of "external actions"; they are not really relevant to the issues I want to emphasize, even though they are essential for a complete mathematical foundation of continuum physics.

Notations: The symbol := indicates that the left side is by definition equal to the right side. The collection of all subsets of a given set $\mathcal{B}$ (the "power set" of $\mathcal{B}$) is denoted by Sub $\mathcal{B}$. The domain, codomain, and range of a given mapping f are denoted by Dom f, Cod f, and Rng f, respectively. The image mapping $f_>$: SubDom f $\longrightarrow$ SubCod f of the mapping f is defined by $f_>(S) := \{f(x) | x \in S\}$ for all $S \in$ SubDom f. The symbols Clo, Int, and Bdy denote the operations of taking the closure, the interior, and the boundary, respectively, of a subset of a topological space. If the domain of a mapping f is a Cartesian product $\mathcal{A} \times \mathcal{B}$ and if $b \in \mathcal{B}$, then $f(\cdot,b) : \mathcal{A} \longrightarrow \operatorname{Cod} f$ is defined by $f(\cdot,b)(x) := f(x,b)$ for all $x \in \mathcal{A}$. The set of all strictly positive real numbers is denoted by $\mathbb{P}^{\times}$. The inverse of an invertible mapping f is denoted by $f^{\leftarrow}$.

2. Material universes

By a material universe we mean a set Ω, endowed with a mathematical structure by the prescription of a relation $\prec$ in Ω. The elements of Ω are to be interpreted as the *subbodies* of a physical body under investigation and the statement $\mathcal{P} \prec \mathcal{Q}$, for any given $\mathcal{P},\mathcal{Q} \in \Omega$, should be interpreted to mean that $\mathcal{P}$ *is a part* of $\mathcal{Q}$.

It is assumed that the structured set Ω satisfies the axioms (M1)-(M6) below. These axioms merely reflect our common sense concerning bodies and their parts.

(M1). $\prec$ is transitive, i.e.,

$$(\mathcal{P} \prec \mathcal{Q} \quad \text{and} \quad \mathcal{Q} \prec \mathcal{R}) \Rightarrow \mathcal{P} \prec \mathcal{R}$$

for all $\mathcal{P},\mathcal{Q},\mathcal{R} \in \Omega$.

(M2). $\prec$ is antisymmetric and reflexive, i.e.,

$$(\mathcal{P} \prec \mathcal{Q} \quad \text{and} \quad \mathcal{Q} \prec \mathcal{P}) \Leftrightarrow \mathcal{P} = \mathcal{Q}$$

for all $\mathcal{P},\mathcal{Q} \in \Omega$.

These first two axioms state that $\prec$ is a (partial) order on Ω. Hence all the concepts and results of the theory of ordered sets apply. In particular, a subset of Ω may or may not have a supremum or infimum. If $\mathcal{P},\mathcal{Q} \in \Omega$, then the supremum [infimum] of $\{\mathcal{P},\mathcal{Q}\}$, if it exists, will be denoted by $\mathcal{P} \vee \mathcal{Q}$ [$\mathcal{P} \wedge \mathcal{Q}$] and called the *join* [*meet*] of $\mathcal{P}$ and $\mathcal{Q}$. (In the past, I have used the term "least envelope" ["greatest common part"].)

(M3). Ω has a maximum, denoted by $\mathcal{B}$, and minimum, denoted by ϕ, so that

$$\phi \prec \mathcal{P} \prec \mathcal{B} \quad \text{for all} \quad \mathcal{P} \in \Omega .$$

The element $\mathcal{B}$ of Ω represents the *whole body* being investigated. The element ϕ of Ω has no physical interpretation. It is useful to have it available in order to simplify the mathematics. We call ϕ the material nothing.

If $\mathcal{P},\mathcal{Q}$ have only ϕ in common, i.e., if $\mathcal{P} \wedge \mathcal{Q} = \phi$, we say that $\mathcal{P}$ and $\mathcal{Q}$ are separate subbodies. We denote the set of all separate pairs of subbodies by

(1.1) $$(\Omega \times \Omega)_{sep} := \{(\mathcal{P},\mathcal{Q}) \in \Omega \times \Omega \mid \mathcal{P} \wedge \mathcal{Q} = \phi\} .$$

(M4). For every $\mathcal{P} \in \Omega$ there is exactly one $\mathcal{Q} \in \Omega$ such that $\mathcal{Q} \wedge \mathcal{P} = \phi$ and $\mathcal{Q} \vee \mathcal{P} = \mathcal{B}$. We denote this $\mathcal{Q}$ by $\mathcal{P}^b := \mathcal{Q}$ and call it the exterior in $\mathcal{B}$ of $\mathcal{P}$.

The subbody $\mathcal{P}^b$ is to be interpreted as the remainder after the subbody $\mathcal{P}$ has been removed from $\mathcal{B}$.

(M5). For every $\mathcal{P},\mathcal{Q} \in \Omega$, if $\mathcal{P} \wedge \mathcal{Q}^b = \phi$ then $\mathcal{P} \prec \mathcal{Q}$.

This axiom is immediately justified by common sense.

(M6). Any subbodies $\mathcal{P},\mathcal{Q} \in \Omega$ have a meet $\mathcal{P} \wedge \mathcal{Q}$.

One can prove from (M1)-(M6) that the operations $\wedge$ and $\vee$ endow $\mathcal{B}$ with the structure of a Boolean algebra. Thus we could have simply assumed that a material universe Ω is a Boolean algebra. However, the relation $\prec$ is closer to physical intuition than the operations $\wedge$ and $\vee$ and the axioms (M1)-(M6) above are more easily justified by common sense than the axioms of a Boolean algebra. Moreover, as we shall see later, there may be circumstances in which one might want to replace (M6) by a weaker axiom.

For any given subbody $\mathcal{P} \in \Omega$, we denote the set of a parts of $\mathcal{P}$ by

$$\Omega_{\mathcal{P}} := \{\mathcal{Q} \in \Omega \mid \mathcal{Q} \prec \mathcal{P}\} .$$

It is easily seen that $\Omega_{\mathcal{P}}$ has itself the natural structure of a material universe. The "is a part of"-relation in $\Omega_{\mathcal{P}}$ is simply the restriction of $\prec$ to $\Omega_{\mathcal{P}}$. The whole body of $\Omega_{\mathcal{P}}$ (axiom (M3)) is $\mathcal{P}$ rather than $\mathcal{B}$ and the exterior in $\mathcal{P}$ (axiom (M4)) of a subbody $\mathcal{R} \in \Omega_{\mathcal{P}}$ is $\mathcal{R}^b \wedge \mathcal{P}$ rather than $\mathcal{R}^b$. The set $\Omega_{\mathcal{P}}$ is also the set of all subbodies that are separate from $\mathcal{P}^b$, i.e.,

(1.2) $$\Omega_{\mathcal{P}} = \{\mathcal{R} \in \Omega \mid (\mathcal{R},\mathcal{P}^b) \in (\Omega \times \Omega)_{sep}\} .$$

3. Interactions

In this section, we assume that a material universe Ω and a linear space $\mathcal{U}$ are given.

We say that a mapping $F: \Omega \longrightarrow \mathcal{U}$ is *additive* if

$$(3.1) \qquad F(\mathcal{P} \vee \mathcal{D}) = F(\mathcal{P}) + F(\mathcal{D})$$

for all $(\mathcal{P},\mathcal{D}) \in (\Omega \times \Omega)_{sep}$.
A mapping $I: (\Omega \times \Omega)_{sep} \longrightarrow \mathcal{U}$ is called an *interaction* in Ω if, for every $\mathcal{P} \in \Omega$, the mappings

$$I(\cdot,\mathcal{P}^b): \Omega_{\mathcal{P}} \longrightarrow \mathcal{U},$$
$$I(\mathcal{P}^b,\cdot): \Omega_{\mathcal{P}} \longrightarrow \mathcal{U}$$

are additive. We say that the interaction I is *skew* if

$$(3.2) \qquad I(\mathcal{P},\mathcal{D}) = -I(\mathcal{D},\mathcal{P})$$

for all $(\mathcal{P},\mathcal{D}) \in (\Omega \times \Omega)_{sep}$. We say that the interaction I is *balanced* if

$$I(\mathcal{P},\mathcal{P}^b) = 0$$

for all $\mathcal{P} \in \Omega$.

Intuitively, it is easiest to interpret an interaction as a system of internal forces. Thus, $I(\mathcal{P},\mathcal{D})$ would be the force exerted on the body $\mathcal{P}$ by the body $\mathcal{D}$. The additivity requirements merely reflect basic prejudices concerning forces. To say that I is a skew interaction expresses the assumption that the forces satisfy the "law of action and reaction". A balanced interaction corresponds to a force system in which the "resultant internal force" on each subbody is zero.

Another physical application of the concept of an interaction occurs in thermodynamics. Here, $I(\mathcal{P},\mathcal{D})$ would be the heat flux from $\mathcal{D}$ into $\mathcal{P}$.

The following easily proved yet important theorem characterizes skew interactions:

Theorem: *The interaction $I: (\Omega \times \Omega)_{sep} \longrightarrow \mathcal{U}$ is skew, i.e., satisfies (3.2), if and only if the mapping*

$$(\mathcal{P} \longmapsto I(\mathcal{P},\mathcal{P}^b)): \Omega \longrightarrow \mathcal{U}$$

is additive.

Corollary: Every balanced interaction is skew.

4. Continuous bodies

By a continuous body we mean a set $\mathcal{B}$ endowed with structure by the prescription of a class $\mathbb{P}$ of mappings. The elements of $\mathcal{B}$ are to be interpreted as material points and the mappings in $\mathbb{P}$ as the possible placements of $\mathcal{B}$ against a suitable background, or frame of reference.

It is assumed that the structured set $\mathcal{B}$ satisfies the axioms (B1)-(B4) below. The axioms (B1)-(B3) depend on the specification of a suitable class $\mathbb{D}$ of mappings called displacements. For our purposes we take $\mathbb{D}$ to consist of all invertible mappings obtained from C^2-diffeomorphisms between Euclidean spaces by restricting both domain and codomain to suitable subsets.

(B1). Each member $\varkappa$ of $\mathbb{P}$ is invertible, its domain is $\mathcal{B}$, and its range $\operatorname{Rng}\varkappa$ is a subset of some Euclidean space $\operatorname{Frm}\varkappa$.

The Euclidean space $\operatorname{Frm}\varkappa$ will be called the frame-space of $\varkappa$. It is interpreted to represent the frame of reference used for the placement $\varkappa$. If $X \in \mathcal{B}$ is a material point, then $\varkappa(X) \in \operatorname{Frm}\varkappa$ is the place of X in the placement $\varkappa$. The subset $\operatorname{Rng}\varkappa$ is the region in $\operatorname{Frm}\varkappa$ occupied by the body $\mathcal{B}$ in the placement $\varkappa$.

(B2). If $\varkappa,\gamma \in \mathbb{P}$, then $\varkappa\circ\gamma^{\leftarrow} \in \mathbb{D}$.

This axiom merely states that a change from one placement to another must belong to the prescribed class of displacements.

(B3). If $\varkappa \in \mathbb{P}$, $\lambda \in \mathbb{D}$ and

$$\operatorname{Rng}\varkappa = \operatorname{Dom}\lambda, \text{ then } \lambda\circ\varkappa \in \mathbb{P}.$$

This axiom states that a placement followed by a displacement is again a placement.

Remark: Whenever the domain or codomain of a mapping is assumed to be a subset of a Euclidean space, this Euclidean space should be regarded as part of the specification of the mapping. Thus, the framespace $\operatorname{Frm}\varkappa$ of a placement $\varkappa$ is considered to be prescribed when $\varkappa$ is given. The class $\mathbb{D}$ of displacements is, strictly speaking, the class of morphisms of a category whose objects are pairs $(\mathcal{D},\mathcal{E})$, where $\mathcal{E}$

is a Euclidean space and $\mathcal{D}$ a subset of $\mathcal{E}$. A morphism from $(\mathcal{D},\mathcal{E})$ to $(\mathcal{D}',\mathcal{E}')$ is an invertible mapping from $\mathcal{D}$ onto $\mathcal{D}'$ that can be extended to a C^2-diffeomorphism from $\mathcal{E}$ onto $\mathcal{E}'$.

A continuous body $\mathcal{B}$ in the sense just described carries a natural topology. If $\mathcal{B}$ is endowed with this topology, then all placements become homeomorphisms.

To do continuum physics, one would wish to apply the theory of interactions described in Sect. 3 to continuous bodies. In order to do so, one must construct a material universe Ω of subbodies in the sense of Sect. 2 from a given continuous body $\mathcal{B}$. The universe Ω should be a collection of subsets of $\mathcal{B}$, i.e., $\Omega \subset \text{Sub}\,\mathcal{B}$, and the "is a part of" relation $\prec$ in Ω should reduce to inclusion $\subset$. The collection Ω should contain only those subsets of $\mathcal{B}$ that are in some sense "nice". In order to specify Ω in a natural way, one must specify a suitable class of "nice" subsets of Euclidean spaces. This class $\mathbb{N}$ must be invariant under displacements, i.e., if $\lambda \in \mathbb{D}$ and $\text{Dom}\,\lambda \in \mathbb{N}$ then $\text{Rng}\,\lambda \in \mathbb{N}$. Once $\mathbb{N}$ is specified, we can state the last axiom:

(B4). For every $\kappa \in \mathbb{P}$, we have $\text{Rng}\,\kappa \in \mathbb{N}$.

This axiom insures that bodies can only occupy "nice" regions in frame-spaces.

The material universe Ω of subbodies of $\mathcal{B}$ can now be defined by

(4.1) $$\Omega := \{\mathcal{P} \in \text{Sub}\,\mathcal{B} \,|\, \kappa_{>}(\mathcal{P}) \in \mathbb{N} \text{ for all } \kappa \in \mathbb{P}\}$$

The invariance of $\mathbb{N}$ under displacements insures that for any given placement $\kappa \in \mathbb{P}$, we have

(4.2) $$\Omega = \{\mathcal{P} \in \text{Sub}\,\mathcal{B} \,|\, \kappa_{>}(\mathcal{P}) \in \mathbb{N}\}$$

i.e., that Ω consists of all those subsets of $\mathcal{B}$ whose image under κ is a "nice" subset of the frame-space Frm^{κ}.

As an example, we may take $\mathbb{N}$ to be the class Regcl of all <u>regularly closed</u> subsets of Euclidean spaces (i.e., all closed sets that coincide with the closure of their interior). In that case, Ω as defined by (4.1) consists of all regularly closed subsets of $\mathcal{B}$. The axioms (M1)-(M6) are all satisfied and we have

(4.3) $$\mathcal{P} \wedge \mathcal{Q} = \text{Clo Int}(\mathcal{P} \wedge \mathcal{Q}) ,$$

(4.4) $$\mathcal{P} \vee \mathcal{Q} = \mathcal{P} \cup \mathcal{Q} \ ,$$

(4.5) $$\mathcal{P}^b = \mathrm{Clo}(\mathcal{B} \setminus \mathcal{P}) \ ,$$

for all $\mathcal{P}, \mathcal{Q} \in \Omega$.

Another example for $\mathbb{N}$ is the class Regop of all <u>regularly open</u> subsets of Euclidean spaces (i.e., all open sets that coincide with the interior of their closure. In that case, Ω is the collection of all regularly open subsets of $\mathcal{B}$. The axioms (M1)-(M2) are again satisfied. The rules (4.3)-(4.5) must be replaced by

(4.6) $$\mathcal{P} \wedge \mathcal{Q} = \mathcal{P} \cap \mathcal{Q} \ ,$$

(4.7) $$\mathcal{P} \vee \mathcal{Q} = \mathrm{Int}\ \mathrm{Clo}(\mathcal{P} \cup \mathcal{Q}) \ ,$$

(4.8) $$\mathcal{P}^b = \mathrm{Int}(\mathcal{B} \setminus \mathcal{P}) \ ,$$

for all $\mathcal{P}, \mathcal{Q} \in \Omega$.

It makes no substantive difference whether one uses the class Regcl or Regop , because the process of taking the closure is a natural one-to-one correspondence from Regop onto Regcl , its inverse being the process of taking the interior. In some contexts it is easier to work with the class Regop , but we prefer Regcl here.

Recall that each Euclidean space has a unique volume-measure, which we denote by vol . Every member of Regcl is volume-measurable and, if not empty, its volume is strictly positive.

When using Regcl or a suitable subclass thereof for $\mathbb{N}$ we define the <u>contact</u> of separate subbodies $\mathcal{P}$ and $\mathcal{Q}$ by

(4.9) $$\mathrm{Cnct}(\mathcal{P},\mathcal{Q}) := \mathcal{P} \cap Q \ .$$

Since $\mathcal{P}$ and $\mathcal{Q}$ are separate, we also have

(4.10) $$\mathrm{Cnct}(\mathcal{P},\mathcal{Q}) = (\mathrm{Bdy}\ \mathcal{P}) \cap (\mathrm{Bdy}\ \mathcal{Q}) \ .$$

Unfortunately, if we use Regcl itself for $\mathbb{N}$, it may happen that the image of the contact $\mathrm{Cnct}(\mathcal{P},\mathcal{Q})$ under some (and hence all) placements fails to be a "surface" in any reasonable sense. It is for this reason that the class Regcl is too large to be suitable for a further development of the foundations of continuum physics.

We might try to take for $\mathbb{N}$ the class of all regularly closed sets that have a piecewise C^2 boundary. More precisely, we may define $\mathbb{N}$ to be the class of all sets of the form $\mathrm{Rng}\ \lambda$, where

$\lambda \in \mathbb{D}$ is a displacement having a closed polyhedron as its domain. If we do so, then the images of contacts do indeed become piecewise C^2 surfaces. The universe Ω of subbodies de ined by (4.1) with this choice of $\mathbb{N}$ satisfies the axioms (M1)-(M5), but not axiom (M6). To see this, one only need construct two regularly closed subsets of a Euclidean space, each with a piecewise C^2 boundary, such that their meet, as given by (4.3), does not have a piecewise C^2 boundary. It is fairly easy to do so.

It is possible to modify the theory of interactions described in Sect. 3 in such a way that it applies to a concept of material universe in which the axiom (M6) is replaced by a weaker one. However, such modification is very awkward. It would be much better if a subclass $\mathbb{N}$ of Regcl could be specified in such a way that all members of $\mathbb{N}$ have a "surface-like" boundary, yet the corresponding Ω satisfies all the axioms of Sect. 2, including (M6). More precisely, $\mathbb{N}$ should have the following properties:

(N1). $\mathbb{N}$ is a subclass of Regcl.

(N2). $\mathbb{N}$ is invariant under displacements.

(N3). If $\mathcal{C},\mathcal{D} \in N$ and $\mathcal{C}$ and $\mathcal{B}$ are both subsets of the same Euclidean space, then $\text{Clo Int}(\mathcal{C} \cap \mathcal{D}) \in \mathbb{N}$, $\mathcal{C} \cup \mathcal{D} \in \mathbb{N}$, and $\text{Clo}(\mathcal{C}\setminus\mathcal{D}) \in \mathbb{N}$.

The following conditions are assumed to apply to every set $\mathcal{C} \in \mathbb{N}$. The Euclidean space of which $\mathcal{C}$ is a subset will be denoted by $\mathcal{E}$ and its translation space by $\mathcal{U}$.

(N4). It is possible to define an area-measure are on Bdy $\mathcal{C}$ in such a way that for each $\mathcal{D} \in \mathbb{N}$ with $\mathcal{D} \subset \mathcal{E}$ and $\text{Clo Int}(\mathcal{C} \cap \mathcal{D}) = \phi$, the subset $\mathcal{C} \cap \mathcal{D}$ of Bdy$\mathcal{C}$ is area-measurable.

(N5). It is possible to define an *exterior unit normal* function $\underline{n}_{\mathcal{C}}$: Bdy$\mathcal{C} \longrightarrow \mathcal{U}$, which assigns to each point of Bdy$\mathcal{C}$ the unit vector normal to the "surface" Bdy $\mathcal{C}$ and directed away from $\mathcal{C}$. The values of $\underline{n}_{\mathcal{C}}$ are unique except perhaps on a subset of Bdy$\mathcal{C}$ that has area-measure zero. If f is a continuous function on an area-measurable subset $\mathcal{S}$ of Bdy$\mathcal{C}$, it is possible to form the area-integral $\int_{\mathcal{S}} f\underline{n}_{\mathcal{C}}d(\text{area}) \in \mathcal{U}$.

(N6). If the set $\mathcal{C}$ is bounded then the Integral-Gradient Theorem (often called Green-Gauss Theorem) applies. We mean by this that for every differentiable function f on $\mathcal{C}$ whose gradient ∇f is volume-integrable, we have

(4.11) $$\int \nabla f d(\mathrm{vol}) = \int_{\mathrm{Bdy}\,\mathcal{C}} f n_{\mathcal{C}} d(\mathrm{area})$$

If the class $\mathbb{N}$ satisfies these conditions (N1)-(N6) then the universe Ω of subbodies defined by (4.1) satisfies all of the axioms (M1)-(M6) and the formulas (4.3)-(4.5) and (4.10) remain valid.

For later use we define here the set

(4.12) $$\Omega_{int} := \{\mathcal{P} \in \Omega \,|\, \mathcal{P} \subset \mathrm{Int}\,\mathcal{B}\}$$

of all internal subbodies of $\mathcal{B}$. It is clear that Ω_{int} consist of all those subbodies whose boundary is included in the interior of $\mathcal{B}$. We say that a mapping $F: \Omega_{int} \longrightarrow \mathcal{U}$, whose codomain $\mathcal{U}$ is a finite-dimensional linear space, is volume-bounded if for some (and hence every) placement $\varkappa$ of $\mathcal{B}$ and some (and hence every) norm ν on $\mathcal{U}$ there is a $k \in \mathbb{P}^{\times}$ such that

(4.13) $$(F(\mathcal{P})) \leqq k\,\mathrm{vol}(\varkappa_{>}(\mathcal{P}))$$

holds for every $\mathcal{P} \in \Omega_{int}$.

5. Contact-interactions

We assume now that a class $\mathbb{N}$ of subsets of Euclidean spaces has been specified such that the condition (N1)-(N6) of the previous section hold. We assume, also, that a continuous body $\mathcal{B}$ is given and that Ω is the universe of subbodies of $\mathcal{B}$ defined by (4.1).

We say that an interaction I in Ω is a contact interaction if, for all $(\mathcal{P},\mathcal{Q}) \in (\Omega \times \Omega)_{sep}$,

$$\mathrm{Cnct}(\mathcal{P},\mathcal{Q}) = \emptyset \Rightarrow I(\mathcal{P},\mathcal{Q}) = 0\,.$$

In other words, a contact interaction is one that can give a non-zero value only if the contact of the bodies is non-empty.

Contact interactions are close to our common-sense idea of a force: A body $\mathcal{Q}$ can exert a force on a body $\mathcal{P}$ only if the two are actually in contact. Of course, since Newton's concept of gravitation we have become used to forces that act over a distance, but the natural philosophers of the late 17th and early 18th century had great difficulties with such forces at first. Mathematically, forces acting over a distance are easier to deal with than contact forces, even though the latter are closer to our common sense. If an interaction represents heat fluxes then the assertion that it is a contact interaction means that bodies can exchange heat only when they are in contact, i.e., by conduction. A heat exchange at a distance would physically be interpreted as an exchange by radiation.

From now on we assume that an interaction I in Ω with values in a finite-dimensional linear space $\mathcal{U}$ is given. If κ is a placement of $\mathcal{B}$ we denote the translation space of the frame space $\mathrm{Frm}\,\kappa$ by $\mathcal{V}_\kappa$.

We say that a continuous function $C_\kappa : \mathrm{Rng}\,\kappa \longrightarrow \mathrm{Lin}(\mathcal{V}_\kappa, \mathcal{U})$ is a *contactor* for the interaction I in the placement κ if, for every $(\mathcal{P}, \mathcal{Q}) \in (\Omega \times \Omega)_{\mathrm{spp}}$, we have

$$(5.1) \qquad I(\mathcal{P},\mathcal{Q}) = \int_{\mathcal{S}} (C_\kappa \underline{n})\, d(\mathrm{area}) ,$$

where $\underline{n} : \mathrm{Bdy}(\kappa_>(\mathcal{P})) \longrightarrow \mathcal{V}_\kappa$ is the exterior unit normal described in (N5) of Sect. 4 and where

$$(5.2) \qquad \mathcal{S} := \kappa_>(\mathcal{P}) \cap \kappa_>(\mathcal{Q}) = \kappa_>(\mathrm{Cnct}(\mathcal{P},\mathcal{Q})) ,$$

which is an area measurable subset of $\mathrm{Bdy}(\kappa_>(\mathcal{P}))$ by (N4) of Sect. 4.

The following facts are easily proved:

(I) <u>If I admits a contactor in some placement, it admits a contactor in every placement</u>.

(II) <u>If I admits a contactor, it is a contact interaction</u>.

(III) <u>If I admits a contactor, it is a skew interaction</u> (in the sense defined in Sect. 3).

(IV) <u>If I admits a contactor, it is</u> *area-bounded* <u>in the following sense</u>: For some (and hence every) <u>placement</u> κ of $\mathcal{B}$ and some (and hence every) norm ν on $\mathcal{U}$ there is a $k \in \mathbb{P}^\times$ such that

$$(5.3) \qquad \nu(I(\mathcal{P},\mathcal{Q})) \leq k \,\mathrm{area}\,(\kappa_>(\mathcal{P}) \cap \kappa_>(\mathcal{Q}))$$

holds for all $(\mathcal{P},\mathcal{Q}) \in (\Omega \times \Omega)_{\mathrm{spp}}$.

(V) <u>If I admits a contactor C_κ which is differentiable and has a bounded and integrable divergence</u> $\mathrm{div}\, C_\kappa : \mathrm{Rng}\,\kappa \longrightarrow \mathcal{U}$ <u>then</u> $(\mathcal{P} \longmapsto I(\mathcal{P},\mathcal{P}^b)) : \Omega_{\mathrm{int}} \longrightarrow \mathcal{U}$ <u>is volume-bounded</u> (in the sense defined by (4.13)).

The fact (V) is a consequence of the formula

$$(5.4) \qquad I(\mathcal{P},\mathcal{P}^b) = \int_{\kappa_>(\mathcal{P})} (\mathrm{div}\, C_\kappa)\, d(\mathrm{vol})$$

valid for all $\mathcal{P} \in \Omega_{\mathrm{int}}$. To obtain (5.4), one only need observe that

$\mathrm{Bdy}\,\varkappa_>(\mathcal{P}) = \mathrm{Bdy}\,\varkappa_>(\mathcal{P}^b)$ and hence that (5.1) reduces to

(5.5) $$I(\mathcal{P},\mathcal{P}^b) = \int_{\mathrm{Bdy}\,\varkappa_>(\mathcal{P})} C_\varkappa \underline{n}\, d(\text{area})$$

if $\mathcal{P} \in \Omega_{\text{int}}$. Application of a suitable Divergence Theorem, which is valid due to (N6) of Sect. 4, then yields (5.4).

In the case when I is a system of internal forces, then a contactor $C_\varkappa$ for I is called a <u>stress-tensor field</u>. If $\varkappa$ is the actual placement at a given present time, then $C_\varkappa$ is usually called the <u>Cauchy-stress</u>; if $\varkappa$ is some reference placement, then $C_\varkappa$ is usually called the <u>Piola-Kirchhoff-stress</u>. In the case when I is a system of heat fluxes, $C_\varkappa$ is called the <u>heat-flux vector field</u>.

It is customary to assume, in most situations in continuum physics, that contact interactions admit a contactor. Perhaps the hardest problem of the mathematical foundations of continuum physics is to decide under what conditions such an assumption is justified. A very good justification would be given by the following result, if it were true.

<u>Desired Theorem</u>: <u>If I is a skew area-bounded interaction such that the mapping $\mathcal{P} \longmapsto I(\mathcal{P},\mathcal{P}^b)$ is volume bounded on Ω_{int}, then I admits a contactor</u>. (The term "area-bounded" is defined in Fact II above and the term "volume bounded" at the end of Sect. 4).

The hypotheses of this Desired Theorem are either consequences of the fundamental laws of the relevant branch of physics or they are reasonable regularity conditions.

In order to describe the progress made to date toward a proof of the Desired Theorem, it is useful to introduce the following concept. Let $\varkappa$ be a placement of $\mathcal{B}$. We denote the unit sphere of $\mathcal{V}_\varkappa$ by $\mathrm{Usph}(\mathcal{V}_\varkappa) := \{\underline{n} \in \mathcal{V}_\varkappa \mid \underline{n}\cdot\underline{n} = 1\}$. We say that a mapping $P_\varkappa : (\mathrm{Rng}\,\varkappa) \times \mathrm{Usph}(\mathcal{V}_\varkappa) \longrightarrow \mathcal{U}$ is a <u>proto-contractor</u> for the interaction I in the placement $\varkappa$ if, for every $(\mathcal{P},\mathcal{Q}) \in (\Omega \times \Omega)_{\text{sep}}$, we have

(5.6) $$I(\mathcal{P},\mathcal{Q}) = \int_{\mathcal{S}} P_\varkappa(x,\underline{n}(x))\, d(\text{area}_x) ,$$

where $\mathcal{S}$ is given by (5.2) and $\underline{n}: \mathrm{Bdy}(\varkappa_>(P)) \longrightarrow \mathrm{Usph}(\mathcal{V}_\varkappa)$ is the exterior unit normal described in (N5) of Sect. 4.

Assume now that I is a skew interaction on $\mathcal{B}$ such that $\mathcal{P} \longmapsto I(\mathcal{P},\mathcal{P}^b)$ is volume-bounded on Ω_{int}. The following two results are steps toward a possible proof of the Desired Theorem:

<u>Theorem A</u>: <u>If I admits a proto-contactor $P_\varkappa$ in a given placement $\varkappa$ such that, for each $\underline{n} \in \mathrm{Usph}(\mathcal{V}_\varkappa)$, the mapping</u>

$$P_\varkappa(\cdot,\underline{n}): \text{Rng}\ \varkappa \longrightarrow \mathcal{U}$$

is bounded and continuous, then I admits a contactor $C_\varkappa$ in $\varkappa$ and

$$P_\varkappa(x,\underline{n}) = C_\varkappa(x)\underline{n} \tag{5.7}$$

holds for all $x \in \text{Rng}\ \varkappa$ and all $\underline{n} \in \text{Usph}(\mathcal{U}_\varkappa)$.

Theorem B: If I is area-bounded, then it admits a bounded proto-contactor in every placement.

If we try to put these two results together to give a proof of the Desired Theorem, we see that there is a gap: Theorem B ensures the existence of a bounded proto-contactor, but in order to apply Theorem A, we need also its continuity in the first variable.

The idea that contact interactions should be given by contactors was introduced in the early 19th century by Fresnel and Cauchy (for forces) and Fourier (for heat fluxes). As a justification, Cauchy proved Theorem A above in 1823. The assumption that a proto-contactor exists and has the continuity property needed in Theorem A is often called Cauchy's stress principle. Theorem B was first proved by Noll in 1959.

Several attempts have been made to narrow the gap described above, the most recent by Gurtin and Martins in 1976 [GM]. However, a completely satisfactory resolution of the problem is still outstanding.

6. Conclusion

In order to arrive at a fully satisfactory mathematical foundation of continuum physics, one must address two issues.

The first is the need to specify a suitable class $\mathbb{N}$ of "nice" subsets of Euclidean spaces, a class that has the properties (N1)-(N6) listed in Sect. 4. It has been clear for several years - to most people who have thought about the problem - that a suitable specification of $\mathbb{N}$ can probably be obtained by using the concept of a "set with locally finite perimeter" from geometric integration theory. However, I do not know of any investigator who has considered all of the conditions (N1)-(N6) in detail. It is possible, in fact, that one can satisfy these conditions only after replacing the concepts of "closure" and "interior" by the concepts of "essential closure" and "essential interior" from geometric integration theory. In any case it seems likely that an appropriate theory can be developed fairly easily. What seems more difficult is to cast such a theory into a form understandable to to people who have not learned - and perhaps do not want to learn - the entire machinery of geometric integration theory as it now stands. It would be highly desirable to simplify matters enough so that the theory can be presented in a graduate introductory course on continuum mechanics.

The second issue is more serious than the first. It is the gap between the Desired Theorem and Theorems A and B of Sect. 5. It is possible, in fact likely, that the Desired Theorem becomes valid only when the definition of "contactor" is modified. For example, the assumption that a contactor be continuous may have to be replaced by a weaker one. In addition, it may not be possible to require that (5.1) hold for all pairs $(\mathcal{P},\mathcal{Q}) \in (\Omega \times \Omega)_{spp}$ but only that it hold for "almost all" pairs, in a sense yet to be determined. The definitions of "area-bounded" and "volume-bounded" also may need modification. The available proofs of Theorems A and B are fairly tedious and take up an excessive amount of time to present in an introductory course on continuum mechanics. It would be nice if an elegant proof of the Desired Theorem, suitably modified, could be found. However, it seems that this can be done only with a radical new idea.

References

[GM] Gurtin, M.E. & Martins, L.C., Cauchy's Theorem in Classical Physics. Archive for Rational Mech. Anal. 60, pp. 305-324 (1976).

[N] Noll, W., Lectures on the Foundations of Continuum Mechanics and Thermodynamics. Archive for Rational Mech. Anal. 52, pp. 62-92 (1973).

Acknowledgement: The research leading to this paper was supported by NSF Grant MCS-8102826.

STRUCTURE OF CONTINUUM PHYSICS

William O. Williams
Department of Mathematics
Carnegie-Mellon University
Pittsburgh, Pennsylvania 15213

In a previous paper Noll has presented a general overview of some problems involved in the formulation of the theory of continuum mechanics; in this paper I describe in more detail the constructions involved in that theory so as to describe what has been done and to indicate the strictures of the unsolved problems. I first describe what I call the 'classical' theory, which is based on classical methods of analysis. The results which I present are due to Gurtin, Martins, Mizel, Noll, and me (in various combinations), but I will not make precise attributions, referring the reader instead to the reference list. Following this I describe the 'modern' theory which (when it is constructed) will utilize more modern notions of analysis. This part must be regarded as speculative; the few specific results described are unpublished.

The Classical Theory. The general problem which we attack is to proceed from a global law of physics, formulated for a material body directly accessible to measurement, to a set of field equations, generally partial-differential equations, whose satisfaction throughout the region occupied by the body guarantees satisfaction of the global law. We will illustrate the process by taking the simplest non-trivial case, energy conservation in a motionless medium. Thus if we denote the region by $\mathcal{B}$ we may state the law of balance of energy in the absence of mechanical working as

1° the rate of increase of the energy stored in $\mathcal{B}$ is equal the net rate of influx of heat into $\mathcal{B}$ from its environment,

and we wish to deduce the equations of heat flow:

2° $\dot{e}(x,t) = \operatorname{div} \underset{\sim}{q}(\underset{\sim}{x},t)$ for all $\underset{\sim}{x}$ in $\mathcal{B}$, all times t,
$\underset{\sim}{q}(x,t)\cdot\underset{\sim}{n} = q_0(t)$ for all $\underset{\sim}{x}$ on $\partial\mathcal{B}$, all times t,

where e is the density of stored energy, $\underset{\sim}{q}$ the heat flux vector, and q_0 the influx of heat through the boundary $\partial\mathcal{B}$. Cauchy apparently was the first to recognize the idea central to such attempts: even to pose the problem of reduction to field equations we are forced to assume that the basic law 1° applies not only to $\mathcal{B}$ but also to any part inside $\mathcal{B}$, taking the environment of that part to include its complement in $\mathcal{B}$.

Accordingly, we are led first to introduce the material universe as a collection Ω of subbodies of $\mathcal{B}$ and an element $\mathcal{B}^e$ which is the environment of $\mathcal{B}$. We presume Ω to have the structure of a Boolean algebra with unit $\mathcal{B}$, as discussed below, and formally extend it to an algebra Ω^e by appending $\mathcal{B}^e$ as a complement of $\mathcal{B}$. We then write $\mathcal{A}^e$ for the join of $\mathcal{B}^e$ and the complement of $\mathcal{A}$ in $\mathcal{B}$. We say that elements in Ω^e are separate if their meet is the null and write $(\Omega\times\Omega^e)_{sep}$ for the collection of separate pairs in $\Omega\times\Omega^e$. Next we introduce the internal energy rate

$$\dot{E} : \Omega \to \mathbb{R}$$

and the heat flux

$$H : (\Omega\times\Omega^e)_{sep} \to \mathbb{R} .$$

A function such as H is said to be an interaction if

$$H(\mathcal{A}\vee\mathcal{C},\mathcal{D}) = H(\mathcal{A},\mathcal{D}) + H(\mathcal{C},\mathcal{D})$$

and

$$H(\mathcal{A},\mathcal{C}\vee\mathcal{D}) = H(\mathcal{A},\mathcal{C}) + H(\mathcal{A},\mathcal{D})$$

for all mutually separate $\mathcal{A},\mathcal{C},\mathcal{D}$.

Assumption. *H is an interaction.*

Finally we formulate our law of balance of energy.

Axiom. *For all* $\mathcal{A} \in \Omega$

$$\dot{E}(\mathcal{A}) = H(\mathcal{A},\mathcal{A}^e) .$$

We assume that elements of Ω are regular closed sets and that the operations in Ω are

$$\mathcal{A} \vee \mathcal{C} = \mathcal{A} \cup \mathcal{C}$$

$$\mathcal{A} \wedge \mathcal{C} = \mathrm{cl}\ \mathrm{int}(\mathcal{A}\cap\mathcal{C}) .$$

We define a contact surface to be a set $\mathcal{S}$ of the form

$$\mathcal{S} = \mathcal{A} \cap \mathcal{C} ,$$

where $(\mathcal{G},\mathcal{C}) \in (\Omega \times \Omega)_{sep}$. Further properties of sets in Ω which are used in the usual proofs of our results are

(a) Ω generates the σ-algebra of Borel subsets of $\mathcal{B}$,

(b) for any $\mathcal{G} \in \Omega$, almost everywhere on $\partial\mathcal{G}$ there exists an exterior normal $\underset{\sim}{n}_{\mathcal{G}}$, and $\mathcal{G}$ admits a Green-Gauss theorem for smooth functions,

(c) if $\mathcal{S}$ is a contact surface, with $\mathcal{S} = \mathcal{G} \cap \mathcal{C}$, then almost everywhere on $\mathcal{S}$

$$\underset{\sim}{n}_{\mathcal{G}} = -\underset{\sim}{n}_{\mathcal{C}} ,$$

(d) if $\mathcal{S}$ is a contact surface, with $\mathcal{S} = \mathcal{G} \cap \mathcal{C}$, then there exist $\hat{\mathcal{G}} < \mathcal{G}$, $\hat{\mathcal{C}} < \mathcal{C}$ of arbitrarily small volume for which $\mathcal{S} = \hat{\mathcal{G}} \cap \hat{\mathcal{C}}$,

(e) if $\mathcal{G},\mathcal{C},\mathcal{D} \in \Omega^e$ are mutually separate then $\mathcal{G} \cap \mathcal{C} \cap \mathcal{D}$ has zero area,

(f) if at a point of tangency $\underset{\sim}{n}_{\mathcal{G}} = \underset{\sim}{n}_{\mathcal{C}}$ then $\partial((\mathcal{C}\backslash\mathcal{G})\cap B_r)\backslash(\partial\mathcal{G}\cup\partial\mathcal{C})$ has area $o(r^2)$; B_r is the ball of radius r centered at that point.

These requirements of smoothness restrict Ω severely and the only collection Ω demonstrated to satisfy all of them is a collection of polytopes. If a more general class is required one may drop the requirement that Ω be a Boolean algebra, i.e., drop the closure assumptions, without affecting any results.

We next specify the nature of the material with which we deal by restricting the behavior of $\dot{E}, H$. We let V denote volume measure, A denote area measure.

<u>Assumption</u>: <u>There exists</u> $k \in \mathbb{R}$ <u>such that</u>

(a) $|\dot{E}(\mathcal{G})| \le kV(\mathcal{G})$ <u>for all</u> $\mathcal{G} \in \Omega$;

(b) $|H(\mathcal{G},\mathcal{C})| \le kA(\mathcal{G}\cap\mathcal{C})$ <u>for all</u> $(\mathcal{G},\mathcal{C}) \in (\Omega\times\Omega)_{sep}$,

(c) $|H(\mathcal{G},\mathcal{B}^e)| \le k(A(\mathcal{G}\cap\partial\mathcal{B}) + V(\mathcal{G}))$ <u>for all</u> $\mathcal{G} \in \Omega$.

Note that we have not assumed that $\dot{E}$ is additive; indeed in many situations in physics one deals with an energy function which is not additive. The question of additivity is tied up with the question of skewness of H .

<u>Proposition.</u> <u>For all separate</u> $\mathcal{G},\mathcal{C} \in \Omega$

$$\dot{E}(\mathcal{G}\vee\mathcal{C}) - \dot{E}(\mathcal{G}) - \dot{E}(\mathcal{C}) = -H(\mathcal{G},\mathcal{C}) - H(\mathcal{C},\mathcal{G})$$

<u>and hence</u> $\dot{E}$ <u>is additive if and only if</u> H <u>is skew.</u>

In order to state our next theorem concisely let us agree to use the term surface flux for a function Q^* which assigns to each contact surface $\mathcal{S}$ and choice of continuous unit normal field $\underset{\sim}{n}$ on $\mathcal{S}$ a number $Q^*(\mathcal{S},\underset{\sim}{n})$.

Theorem (properties of the heat flux).

(i) H is skew (and hence $\dot{E}$ is additive).

(ii) H reduces to a surface flux in $\mathcal{B}$; there exists a surface flux Q^* such that

$$H(\mathcal{A},\mathcal{C}) = Q^*(\mathcal{S},\underset{\sim}{n})$$

whenever $\mathcal{A}$ and $\mathcal{C}$ are separate, $\mathcal{S} = \mathcal{A} \cap \mathcal{C}$, and $\underset{\sim}{n}$ is the outward unit normal to $\mathcal{A}$ on $\mathcal{S}$.

Let $\mathcal{D}_\rho(\underset{\sim}{x})$ denote a plane disc of radius ρ centered at $\underset{\sim}{x}$ and let $\underset{\sim}{n}$ be normal to $\mathcal{D}_\rho(\underset{\sim}{x})$. Defining

$$q_\rho(\underset{\sim}{x},\underset{\sim}{n}) = \frac{Q^*(\mathcal{D}_\rho(\underset{\sim}{x}),\underset{\sim}{n})}{\pi\rho^2} ,$$

it is clear that $\lim_{\rho\to 0} q_\rho(\underset{\sim}{x},\underset{\sim}{n})$ exists for almost every $\underset{\sim}{x}$ in any plane $\mathcal{P}$ with normal $\underset{\sim}{n}$. The following assumption strengthens this result.

Assumption. Q^* has uniform density: for each $\underset{\sim}{n}$,

$$\lim_{\rho\to 0} q_\rho(\cdot,\underset{\sim}{n})$$

exists uniformly in $\overset{\circ}{\mathcal{B}}$.

The great failing of the classical theory is the inability to evade this assumption, which is difficult to justify from any global arguments; it is equivalent, as we now see, to an assumption of continuity.

Theorem.

(i) There exists a field $\dot{e} \in L^\infty(\mathcal{B})$ such that for all $\mathcal{A} \in \Omega$,

$$\dot{E}(\mathcal{A}) = \int_{\mathcal{A}} \dot{e}\, dV .$$

(ii) There exists a vector field $\underset{\sim}{q} \in C(\mathcal{B})$ such that if $\mathcal{A}$ and $\mathcal{C}$ are separate subbodies,

$$H(\mathcal{A},\mathcal{C}) = \int_{\mathcal{S}} \underset{\sim}{q}\cdot\underset{\sim}{n}\, dA ,$$

where $\mathcal{S} = \mathcal{G} \cap \mathcal{C}$ and n is the outward unit normal to $\mathcal{G}$.

(iii) There exists a field $r \in L^\infty(\mathcal{B})$ such that for any subbody $\mathcal{G}$,

$$H(\mathcal{G},\mathcal{B}^e) = \int_{\mathcal{G}\cap\partial\mathcal{B}} \underset{\sim}{q}\cdot\underset{\sim}{n}\, dA + \int_{\mathcal{G}} r\, dV .$$

These representation theorems suffice to deduce the local balance law. We say that $\underset{\sim}{q}$ has a weak divergence at x if

$$\lim_{n\to\infty} \frac{1}{V(\mathcal{G}_n)} \int_{\partial\mathcal{G}_n} \underset{\sim}{q}\cdot\underset{\sim}{n}\, dA$$

exists for any sufficiently regular sequence of subbodies shrinking to $\{x\}$.

Theorem (local form of the first law). $\underset{\sim}{q}$ has a weak divergence and

$$\dot{e} = \operatorname{div} \underset{\sim}{q} + r$$

almost everywhere in $\mathcal{B}$, while almost everywhere on $\partial\mathcal{B}$

$$\underset{\sim}{q}\cdot\underset{\sim}{n} = q_0 .$$

Here q_0 is the density of $H(\cdot,\mathcal{B}^e)$ on $\partial\mathcal{B}$.

In fact it is easy to see that these local relations, with the representations above, are equivalent to the axioms on energy.

The Weak Theory. In modern analysis one requires the balance equation not in the form of a partial differential equation but in a weak form. Thus the goal of the weak theory is to produce an equation

$$\int_{\mathcal{B}} \dot{e}\varphi\, dV = \int_{\mathcal{B}} (\nabla\varphi\cdot\underset{\sim}{q}+\varphi r)\, dV + \int_{\partial\mathcal{B}} (q_0\varphi)\, dA \qquad (*)$$

valid for any test function φ in $\mathcal{B}$. In fact the classical formulation is equivalent to (*): Antman & Osborn have shown that equivalence of (*) to

$$\int_{\mathcal{G}} \dot{e}\, dV = \int_{\partial\mathcal{G}} \underset{\sim}{q}\cdot\underset{\sim}{n}\, dA + \int_{\mathcal{G}} r \quad \text{for all } \mathcal{G} \in \Omega$$

$$\underset{\sim}{q}\cdot\underset{\sim}{n} = q_0 \ \text{A-a.e. on } \partial\mathcal{B}$$

requires only that $\dot{e}$, $\underset{\sim}{q}$ and r be in $\mathcal{L}_{\alpha,\mathrm{loc}}(\mathcal{B})$ for some $\alpha \geq 1$ (and q_0 be in the corresponding Sobelev space on $\partial\mathcal{B}$). These conditions certainly are included in our conclusions, so that the weak equation obtains. However, leaving aside even the problem of the distasteful assumption of uniformity on Q^*, it is much to be preferred that one formulate the entire theory in terms which aim toward (*) as the final outcome and which do not require classical constructions and techniques. Only a small amount of progress has been made in this direction.

To generalize the class of subbodies the obvious choice is to consider the collection of sets of finite perimeter, as this is the broadest collection which admits a notion of measurable surfaces and a version of the Green-Gauss theorem. It is not difficult to show that this collection forms a ring of sets. However it is true that operations with this class of sets involve an identification of sets V-almost everywhere, so it is useful to restrict the class considered. One construction, obviously analogous to the classical choice, is to consider in Ω only those sets of finite perimeter which are * regular * closed sets:

$$\mathcal{A}_*^{\ *} = \mathcal{A} .$$

Here $\mathcal{A}_*$ denotes the set of points of density of $\mathcal{A}$, $\mathcal{C}^*$ the complement of the set of points of density of the complement of $\mathcal{C}$. One can show that this collection is a Boolean algebra under the operations

$$\mathcal{A} \vee \mathcal{C} = \mathcal{A} \cup \mathcal{C}, \quad \mathcal{A} \wedge \mathcal{C} = (\mathcal{A} \cap \mathcal{C})_*^{\ *} .$$

Moreover it is possible to show that this collection obeys the rules (a)-(e) previously required of Ω; it is not clear whether (f) also is true, but it is possible to evade this requirement, used in the proof that H reduces to a surface flux, by using approximation results (I shall establish these assertions in a work now in preparation).[1] It then follows that the classical arguments may be applied in essentially unaltered form to deduce a local balance equation.

But this falls well short of the desiderata. In the first place, borrowing of these classical arguments leaves us with the difficulty of the unrealistic assumption of smoothness. A more serious objection is the second: taking this sort of detour into classical arguments and smooth densities, only to return to the weak form by the Antman-Osborn result, is completely foreign to the nature of the hypotheses and of the conclusion. We lack still a satisfactory direct argument.

---

[1] M. Gurtin has informed me that William Ziemer has recently established that the reduction to a surface flux is valid for all sets of finite perimeter; in my construction I found it necessary to limit the domain of surface fluxes to include only locally rectifiable surfaces.

To attempt construction of a satisfactory theory the starting place is clearly the theory of functions of bounded variation and of geometric measure theory. Unfortunately it does not seem to be possible directly to borrow results from the latter theory. To illustrate, consider the central problem of reducing H . One natural line of procedure would be as follows. We wish to establish the existence of a linear form Q^* on an appropriate set of measures so that

$$H(\mathcal{G},\mathcal{C}) = \langle Q^*, \varphi_{\mathcal{C}} \nabla \varphi_{\mathcal{G}} \rangle ,$$

where $\varphi_{\mathcal{G}}$ denotes the support function of the set $\mathcal{G}$. Then one could presumably use standard denseness arguments to establish that the values of

$$H(\mathcal{G},\mathcal{B}\setminus\mathcal{G}) = \langle Q^*, \nabla\varphi_{\mathcal{G}} \rangle$$

are uniquely determined by values

$$\langle Q^*, \nabla\varphi \rangle$$

for test functions φ . This sort of result then would lead to weak forms of the equations. But what is the nature of Q^* ? The natural duality with which geometric measure theory works is between C^∞ 2-forms and 2-currents, but we find that the inherent smoothness conditions defeat attempts to apply these notions to the above construction (or to the obvious dual construction). For such reasons I feel that the advance of our continuum mechanics structure must await the construct of some theory analogous to that of geometric measure theory but with a more natural duality.

Acknowledgment. The work underlying this paper was supported by a grant from the National Science Foundation.

References

Noll, W., The foundations of classical mechanics in the light of recent advances in continuum mechanics. In: The Axiomatic Method with Special Reference to Geometry and Physics, 266-281, Amsterdam, North Holland Co., 1959.

Gurtin, M. E. and W. O. Williams, An axiomatic foundation for continuum thermodynamics, Arch. Rational Mech. Anal., 26, 83-117, 1967.

Gurtin, M., V. Mizel, and W. Williams, On Cauchy's stress theorem, J. Math. Anal. Appl., 22, 398-401, 1968.

Williams, W. O., On internal interactions and the concept of thermal isolation, Arch. Rational Mech. Anal., 34, 245-258.

Federer, H., Geometric Measure Theory, New York, Springer-Verlag, 1969.

Gurtin, M. E. and W. O. Williams, On the first law of thermodynamics, Arch. Rational Mech. Anal., 42, 77-92, 1971.

Gurtin, Morton E. and Luiz C. Martins, Cauchy's theorem in classical physics, Arch. Rational Mech. Anal., 60, 305-324, 1976.

Antman, S. and J. E. Osborn, The prin iple of virtual work and integral laws of motion, Arch. Rational Mech. Anal., 69, 231-262, 1979.

Williams, W., On a class of subbodies for continuum mechanics, in preparation.

Reference appended by Editor:

Ziemer, W. P., Cauchy Flux and Sets of Finite Perimeter, Archive for Rational Mechanics and Analysis 84, 189-201, 1983.

ON DIFFERENTIABLE SPACES[1]

Kuo-Tsai Chen
Department of Mathematics
University of Illinois
Urbana, IL 61801

The purpose of our notion of differentiable spaces is to provide a formal framework for differential and integral calculus. We start our constructions from an appropriately chosen model category, where usual notions of calculus are well defined. This is actually the very approach used implicitly in classical analysis. Particularly in our mind is the classical theory of calculus of variations. In this case, the model category consists of a single object, namely, an interval, which, for simplicity, is taken to be the unit interval I. The morphisms are all C^∞ maps $I \longrightarrow I$. A path in an open set M of R^n is a C^∞ map $I \longrightarrow M$. The differentiable structure of the path space $P(M)$ of M is determined by variations, which are maps of the type $\alpha: I \longrightarrow P(M)$ such that the evaluation map $I \times I \longrightarrow M$ given by $(s,t) \longrightarrow \alpha(s)(t)$ is a C^∞ map. Critical points of a function on $P(M)$ can then be defined through variations by pulling back to the parameter space I. In order to consider the second variation, the model category is later enlarged to include also $I \times I$ as an object.

1. We take as the model category the one whose objects are convex subsets with nonempty interior in R^n, $n = 0,1,\ldots$, and whose morphisms are C^∞ maps. If U is such a convex set, then U is equipped with coordinates $\xi = (\xi^1,\ldots,\xi^n)$, namely, those of R^n. Moreover, owing to the convexity, partial derivatives of a function on U can be defined everywhere up to an arbitrary order.

By a convex set, we shall always mean a convex set with nonempty interior in R^n as described above, where n can be an arbitrary natural integer.

Definition 1.1. A C^∞ space M is a set equipped with a family of set maps called plots, which satisfy the following conditions:

(a) Every plot is a map of the type $U \longrightarrow M$ where U is a convex set.

(b) If $\phi: U \longrightarrow M$ is a plot and if U' is also a convex set

[1] Work supported in part by NSF Grant MCS 82-00775.

(not necessarily of the same dimension as U), then, for every C^∞ map $\theta: U' \longrightarrow U$, $\phi\theta$ is also a plot.

(c) Every constant map from a convex set to M is a plot.

(d) Let $\{U_i\}$ be an open convex covering of a convex set U, and let $\phi: U \longrightarrow M$ be a set map. If each restriction $\phi|U_i$ is a plot, then ϕ itself is a plot.

Definition 1.2. Let M and M' be C^∞ spaces. A set map $f: M' \longrightarrow M$ is a C^∞ map if, for every plot ϕ of M', $f\phi$ is a plot of M.

The de Rham complex functor $\Lambda = \{\Lambda^p, d\}$ is well defined on the model category and can be extended to a functor on the category of C^∞ spaces and C^∞ maps as follows:

A p-form w on a C^∞ space M is a rule that assigns to each plot $\phi: U \longrightarrow M$ a p-form w_ϕ on U satisfying the following compatibility condition: If U' is a convex set and if $\theta: U' \longrightarrow U$ is a C^∞ map, then $w_{\phi\theta} = \theta^* w_\phi$. If w is a p-form on M, then dw is the (p+1)-form on M such that $(dw)_\phi = dw_\phi$ for every plot ϕ of M. The addition and multiplication of forms are defined in the obvious way.

This is, as a matter of fact, a general procedure of extending a functor on the model category to a functor on the category of C^∞ spaces and maps.

Examples of C^∞ spaces abound. We list several of them.

(a) Every C^∞ manifold M is a C^∞ space, whose plots are C^∞ maps.

(b) Every subset of a C^∞ space is naturally a C^∞ space. In particular, an arbitrary subset of R^n is a C^∞ space.

(c) The cartesian product of two C^∞ spaces is naturally a C^∞ space.

(d) Let M and M' be C^∞ spaces. Then the set $C^\infty(M',M)$ of all C^∞ maps $M' \longrightarrow M$ is a C^∞ space. A plot $\alpha: U \longrightarrow C^\infty(M',M)$ is a set map whose evaluation map $U \times M' \longrightarrow M$ is a C^∞ map.

There have been earlier (see [3]) treatments of differentiable spaces. The notion of C^∞ spaces as described above distinguishes itself by not requiring topology. Usefulness of this notion has been demonstrated in applications to calculus of variations and to path space cohomology. We are going to make brief explanation of these two aspects.

2. Classical calculus of variations is the critical point theory of certain type of functions on the path space of a region in R^n. Functions (or functionals, in conventional terminology) in consideration are of the type of a path integral $\int F(x,\dot{x})dt$. A path is a critical point of this function if and if the Euler equations

$$\frac{\partial}{\partial x^i} F(x,\dot{x}) - \frac{\partial}{\partial t}\frac{\partial}{\partial \dot{x}^i} F(x,\dot{x}) = 0 , \qquad i = 1,\dots,n .$$

are satisfied along the path. We now proceed to describe how the Euler equations are meaningful on the path space of any C^∞ space. For detail, see [2].

Let M be an arbitrary C^∞ space. Let F be a rule that assigns to each plot $\phi\colon U \longrightarrow M$ a function $f_\phi = F_\phi(\xi,\xi)$. Roughly speaking, F_ϕ is a function on the tangent bundle of U. The collection $\{f_\phi\}$ is required to satisfy the familiar compatibility condition (i.e. the chain rule). Then F gives rise to a function $\int F dt$ on the path space $P(M) = C^\infty(I,M)$. Let $\underline{E}F$ be the rule that assigns to each plot $\phi\colon U \longrightarrow M$ a 1-form

$$(\underline{E}F)_\phi = \sum\left\{\frac{\partial}{\partial \xi^i} F_\phi(\xi,\xi) - D_t \frac{\partial}{\partial \xi^i} F_\phi(\xi,\xi)\right\} d\xi^i ,$$

where D_t is the formal differential operator so that $D_t\xi = \xi$ and $D_t\xi = \ddot{\xi}$. Roughly speaking, $(\underline{E}F)_\phi$ is a 1-form on the second order tangent bundle of U, and the usual compatibility condition holds for the collection $\{(\underline{E}F)_\phi\}$.

Let x_0 and x_1 be given points of M, and let $P(M;x_0,x_1)$ denote the C^∞ subspace of $P(M)$ consisting of all paths from x_0 to x_1. It can be shown that a path is a critical point of the function $\int F dt$ restricted to $P(M;x_0,x_1)$ if and only if $\underline{E}F$ vanishes along the path. This intrinsic formulation not only makes the Euler equations valid on an arbitrary C^∞ space but also facilitates computations.

3. The de Rham complex $\Lambda(P(M))$ of the path space $P(M)$ of a C^∞ space M (or, more concretely, a C^∞ manifold) is extremely complicated. We narrow our attention to a subcomplex $\Lambda(M)'$ constructed from the de Rham complex $\Lambda(M)$ in the following manner:

Let $p_0, p_1\colon P(M) \longrightarrow M$ be the end point map given by $p_0(\gamma) \longmapsto \gamma(0)$ and $p_1(\gamma) \longmapsto \gamma(1)$ for every path $\gamma\colon I \longrightarrow M$. Then p_0 and p_1 are C^∞ maps and induce respectively differential graded homomorphisms

$$p_0^*,\ p_1^*:\ \Lambda(M) \longrightarrow \Lambda(P(M)) .$$

Let $\eta: M \longrightarrow P(M)$ be the C^∞ map sending $x \in M$ to the constant path at x. There is a canonical homotopy G from the C^∞ map ηp_0 to the identity map of $P(M)$. This C^∞ homotopy is carried out by contracting each path along itself to its initial point. The induced cochain homotopy

$$\int_G : \Lambda(P(M)) \longrightarrow \Lambda(P(M))$$

is a graded map of degree -1 such that

$$\int_G du + d\int_G u = u - (\eta p_0)^* u , \qquad u \in \Lambda(P(M)) .$$

We define $\Lambda(M)'$ to be the smallest subalgebra of $\Lambda(P(M))$ generated by $p_0^*(M)$ and $p_1^*(M)$ and stable under $\int_G$. It turns out that $\Lambda(M)'$ is a subcomplex of $\Lambda(P(M))$ and is spanned by elements of the type

$$p_0^* w' \wedge (\int w_1 \ldots w_r) \wedge p_1^* w''$$

where $w', w'', w_1, \ldots, w_r$ are forms on M and, inductively, $w_1 = \int_G p_1^* w_1$ and, for $r > 1$,

$$\int w_1 \ldots w_r = \pm \int_G (\int w_1 \ldots w_{r-1}) \wedge p_1^* w_r .$$

Path space differential forms of this type are called iterated (path) integrals. Under reasonable conditions, this complex $\Lambda(M)'$ gives rise to associated complexes having correct cohomology for the loop space and various other spaces related to the path space of a simply connected C^∞ space. For detail, see [1].

4. Our notion of a C^∞ space M can be further generalized by not requiring that M is a set. In this case, a plot is no more a set map, and, to every plot ϕ, there is an associated convex set $U = U(\phi)$. The conditions (c) and (d) in Definition 1.1 need modification. We do not intend to spell out precisely what the generalization should be, but we are going to point out an example showing that such a generalized

notion is perhaps natural and desirable.

Let M be an arbitrary C^∞ space as given by Definition 1.1. We define an extended plot α of $P(M)$ to be a convex set $U = U(\alpha)$ together with a family of plots $\alpha_i: U_i \longrightarrow P(M)$ such that $\{U_i\}$ is an open covering of U and that, for every $u \in \Lambda(M)'$, u_{α_i} and u_{α_j} coincide on the intersection $U_i \cap U_j$ for every pair of indices i and j. Then, for every $u \in \Lambda(M)'$, there is a well defined form u_α on U with $u_\alpha | U_i = u_{\alpha_i}$. Two extended plots α and α' of $P(M)$ are equivalent if $U(\alpha) = U(\alpha')$ and, for every $u \in \Lambda(M)'$, $u_\alpha = u_{\alpha'}$.

Let $[\alpha]$ denote the equivalence class of an extended plot α. Then $[\alpha]$ is not a set map, and the domain $U = U(\alpha)$ of $[\alpha]$ is a well defined convex set. Moreover, if $\theta: U' \longrightarrow U$ is a C^∞ map of convex sets and if $\alpha = \{U: \alpha_i\}$ and $\alpha' = \{U; \alpha'_j\}$ are equivalent extended plots, so are $\alpha\theta = \{U'; \alpha_i\theta\}$ and $\alpha'\theta = \{U'; \alpha'_j\theta\}$.

If $\theta: U' \longrightarrow U$ is an inclusion map, we shall write

$$\alpha | U' = \alpha\theta .$$

Let $[P(M)]$ consist of all equivalence classes of extended plots of $P(M)$. Then $[P(M)]$ possesses properties similar respectively to conditions (a), (b) and (d) of Definition 1.1. The first two properties can be easily seen. In relation to the condition (d), we let U be a convex set and let $\{[\alpha_i]\}$ be a family of equivalence classes of extended plots of $P(M)$ of U. If $\{U(\alpha_i)\}$ is an open covering of U and if $[\alpha_i | U_i \cap U_j] = [\alpha_j | U_i \cap U_j]$, then there exists a unique equivalence class $[\alpha]$ of extended plots of $P(M)$ with $U(\alpha) = U$ such that $[\alpha | U_i] = [\alpha_i]$ for each i.

The above example seems to justify a more category-theoretical approach to our notion of C^∞ spaces. It should be meniioned here that there are other recent treatments of differentiable spaces. In particular, we have in mind the treatment of A. Kock and others on synthetic differential geometry.

Bibliography

1. K. T. Chen, Iterated path integrals, Bull. Amer. Math. Soc. 83 (1977), 831-879.

2. __________, The Euler operator, Arch. Rat. Mech. & Anal. 75 (1981), 175-191.

3. J. W. Smith, The de Rham theorem for general spaces, Tohoku Math. J. (2) 18 (1966), 115-137.

CARTESIAN CLOSED CATEGORIES AND ANALYSIS OF SMOOTH MAPS

Alfred Frölicher
Section de Mathématiques
Université de Genève
2-4, rue du Liévre
CH-1211 GENEVE 24

The purpose of this article is to give a survey of a new and simple approach to the analysis of smooth maps, to indicate its good categorical properties, and to show that it includes classical calculus based on Banach spaces.

We shall equip for classical smooth manifolds V and W (in fact V,W can be more general) the function space $C^{\infty}(V,W)$ with a smooth structure. This is not possible if one restricts to Banach manifolds. Many attempts have been made to generalize the notions "differentiable" and "smooth" from maps between Banach spaces to maps between more general vector spaces (all vector spaces are supposed over $\mathbb{R}$). In infinite dimension, the differentiability of a map between vector spaces depends on some additional structure of the vector spaces, as e.g. a norm, a topology, a pseudotopology, a bornology. Since in the traditional set-up the derivative of a map $f: E_1 \to E_2$ involves the function space $L(E_1,E_2)$, there were problems concerning the existence of a function-space structure with the property that the composite of two twice differentiable functions is again twice differentiable. The best way to overcome these difficulties was by using vector spaces over a cartesian closed category. So pseudotopologies were used in [1], [4], compactly generated topologies in [16], bornologies in [3], arc-wise determined spaces in [10]. We try to give a natural answer to the question: which is the "good" category of vector spaces for which calculus should be developed, and we shall see that the explicit description of these vector spaces can be given in many different ways.

The author wishes to thank A. Kriegl, F. W. Lawvere, L. Nel and S. H. Schanuel for their very helpful suggestions and discussions.

Categories generated by sets of maps

In [5], any monoid M of maps of any fixed set B into itself was used to generate a category $\underline{K}_M$, and a condition on M for $\underline{K}_M$ to be cartesian closed was given. Instead of using a submonoid M of B^B one could take any subset M of B^B, but the category $\underline{K}_M$ depends only on the submonoid generated by M. However, as pointed out by Lawvere, Schanuel and Zame [13] it is useful to start with a set M of maps between two possibly different basic sets A and B, i.e. $M \subset B^A$. $\underline{K}_M$ is also called the Petermann category of M.

A M-structure on a set E is a couple (C,F) where $C \subset E^A$ and $F \subset B^E$ such that for $c : A \longrightarrow E$ resp. $f : E \longrightarrow B$ one has $c \in C$ iff $f \circ c \in M$ for all $f \in F$ and similarly $f \in F$ iff $f \circ c \in M$ for all $c \in C$. The objects of $\underline{K}_M$ are triples (E,C,F) where E is a set and (C,F) a M-structure on E; the morphisms from (E_1,C_1,F_1) to (E_2,C_2,F_2) are those maps $\varphi : E_1 \longrightarrow E_2$ which satisfy $\varphi_*(C_1) \subset C_2$ or equivalently $\varphi^*(F_2) \subset F_1$. All M-structures on a given set E form a complete lattice with respect to the order "finer" (for which one requires the identity map of E to be a morphism). Any set $F_0 \subset B^E$ generates a M-structure (C,F) for which $C = \{c : A \longrightarrow E;$ $f \circ c \in M \; \forall f \in F_0\}$; it is the coarsest satisfiing $F_0 \subset F$. Similarly for any given $C_0 \subset E^A$. In $\underline{K}_M$ all limits and colimits exist, they are obtained explicitly by putting on the (co-)limit of the underlying sets the initial resp. final M-structure.

We shall use one example where $A \neq B$ because it is very useful for investigating the example $M = C^\infty(\mathbb{R},\mathbb{R})$ (cf [13]), namely $A = \mathbb{N}$, $B = \mathbb{R}$, $M = \ell^\infty$ (the set of bounded sequences of $\mathbb{R}$). ℓ^∞-structures are related to Kolmogorov-bornologies (cf [8]). Examples of ℓ^∞-structures are obtained by taking as set E the underlying set of either a metric space or a locally convex space and by defining

$$C = \{c : \mathbb{N} \longrightarrow E;\; c(\mathbb{N}) \text{ is bounded}\},$$

$$F = \{f : E \longrightarrow \mathbb{R};\; f \circ c \in \ell^\infty \;\; \forall c \in C\}.$$

One verifies that (C,F) is a ℓ^∞-structure on E.

For calculus, examples where M consists of certain (differentiable) maps $\mathbb{R} \longrightarrow \mathbb{R}$ have to be investigated, e.g. $C(\mathbb{R},\mathbb{R})$; $C^k(\mathbb{R},\mathbb{R})$; $C^\infty(\mathbb{R},\mathbb{R})$; $\mathrm{Lip}(\mathbb{R},\mathbb{R})$; $C^{k,1}(\mathbb{R},\mathbb{R})$; $B(\mathbb{R},\mathbb{R})$; $B_\ell(\mathbb{R},\mathbb{R})$. Here "Lip" refers to the Lipschitz condition with exponent 1 which has to be satisfied locally; $C^{k,1}$ means k times differentiable and such that the k-th derivative lies in $\mathrm{Lip}(\mathbb{R},\mathbb{R})$; B (resp B_ℓ) means bounded (resp. locally bounded) functions $B(\mathbb{R},\mathbb{R})$ and $B_p(\mathbb{R},\mathbb{R})$ yield the same $\underline{K}_M$ as ℓ^∞.

For $k \in \mathbb{N}$, the category $\underline{K}_M$ generated by $M = C^k(\mathbb{R},\mathbb{R})$ turns out to be not very useful, since for $n > 1$

$$(\mathbb{R}^n, C^k(\mathbb{R},\mathbb{R}^n),\; C^k(\mathbb{R}^n,\mathbb{R}))$$

fails to be an object. However, the above triple is an object for $k = \infty$ and also if one replaces C^k by $C^{k,1}$; this is in fact essentially equivalent to the famous theorem of Boman [2] saying that a real function on $\mathbb{R}^n$ is C^∞ (resp. $C^{k,1}$) if it is C^∞ (resp. $C^{k,1}$) along each smooth curve of $\mathbb{R}^n$. Using partitions of unity one deduces easily that the same holds if $\mathbb{R}^n$ is replaced by a paracompact C^∞-resp. $C^{k,1}$-manifold. It follows that these manifolds form full subcategories of the respective category $\underline{K}_M$. In order to show the same for many infinite dimensional manifolds one uses the generalization of Boman's theorem given in [6]; we restrict to the smooth case.

Theorem 1. For a map $\varphi : E_1 \to E_2$ between Fréchet spaces the following conditions are equivalent:

1. φ is smooth,
2. $\varphi_*(C^\infty(\mathbb{R},E_1)) \subset C^\infty(\mathbb{R},E_2)$,
3. $\varphi^*(E_2') \subset C^\infty(E_1,\mathbb{R})$,
4. $C^\infty(E_2,\mathbb{R}) \circ \varphi \circ C^\infty(\mathbb{R},E_1) \subset C^\infty(\mathbb{R},\mathbb{R})$.

Though there are many notions of differentiability of a map between Fréchet spaces, they almost all yield the same infinitely differentiable maps; this is the meaning of "smooth" in (1), which in particular is the usual one in case of Banach spaces. In (3), E_2' notes the topological dual of E_2.

As corollary one obtains:

Theorem 2. $(V, C^\infty(\mathbb{R},V), C^\infty(V,\mathbb{R}))$ is an object of the category $\underline{C}^\infty$ generated by $C^\infty(\mathbb{R},\mathbb{R})$ for

a) V any Fréchet space

b) V any paracompact C^∞-manifold over a Fréchet space E which has enough smooth functions (i.e. to any 0-neighborhood U there exists $f \in C^\infty(E,\mathbb{R})$ with $f(0) = 1$ and $f(x) = 0$ for $x \notin U$; e.g. if E is nuclear).

It follows that these Fréchet manifolds form a full subcategory of $\underline{C}^\infty$.

Cartesian closedness

We assume that the given set $M \subset B^A$ contains all constant maps $A \to B$. Then the one point set has a unique M-structure yielding an object S which is final in $\underline{K}_M$, hence $S\pi\text{-}\cong\text{Id}$. S yields also a representation of the forgetful functor V from $\underline{K}_M$ into Set: $V \cong \underline{K}_M(S,-)$.

Let us suppose now that the category $\underline{K} = \underline{K}_M$ is cartesian closed. Then there exists a functor $H : \underline{K} \times \underline{K}^{op} \longrightarrow \underline{K}$ such that

(1) $$\underline{K}(X,H(Y,Z)) \cong \underline{K}(X\pi Y,Z) .$$

Hence for $X = S$ one has $VH(Y,Z) \cong \underline{K}(Y,Z)$ and it is easy to show that by modifying H if necessary one can obtain $VH(Y,Z) = \underline{K}(Y,Z)$ and also the bijections in (1) to be of the form $f \longrightarrow \tilde{f}$ where $\tilde{f}(x,y) = f(x)(y)$.

The basic sets A resp. B have natural M-structures of the form (C_A,M) resp. (M,F_B) where C_A and F_B are determined by M (cf. the axioms of a M-structure). We note $\overline{A}$ resp. $\overline{B}$ the objects of $\underline{K}$ so obtained and we remark that for any object $X = (E_X,C_X,F_X)$ one has $C_X = \underline{K}(\overline{A},X)$ and $F_X = \underline{K}(X,\overline{B})$. The (modified) functor H yields the object $H(\overline{A},\overline{B})$; its underlying set is M , and we denote its M-structure by (Γ,Φ) . Since $\Gamma = \underline{K}(\overline{A},H(\overline{A},\overline{B})) \cong \underline{K}(\overline{A}\pi\overline{A},\overline{B})$ we have

(2) $$\Gamma = \{\gamma : A \longrightarrow M;\ \tilde{\gamma}\circ(\sigma,\tau) \in M \quad \forall \sigma,\tau \in C_A\} ;$$

(3) $$\Phi = \{\varphi : M \longrightarrow B;\ \varphi\circ\gamma \in M \quad \forall\gamma \in \Gamma .$$

Hence cartesian closedness of $\underline{K}_M$ implies the following condition on M (Γ and Φ being determined by (2) resp. (3)):

(4) $$\left.\begin{array}{l} \gamma : A \longrightarrow M \\ \varphi\circ\gamma \in M \ \forall\varphi \in \Phi \end{array}\right\} \Rightarrow \gamma \in \Gamma$$

One easily shows that the converse holds; hence

<u>Theorem 3</u>. $\underline{K}_M$ is cartesian closed if and only if M satisfies (4).

The verification of (4) is not difficult for $M = C(\mathbb{R},\mathbb{R})$; in this case $\underline{K}_M$ is isomorphic to the cartesian closed category of imbeddable arc-wise determined spaces. (4) also holds for $M = \ell^\infty$; the verification is almost trivial.

The case $M = C^\infty(\mathbb{R},\mathbb{R})$ is more difficult and justifies some further comment. By Boman's theorem one gets in this case

$$\Gamma = \{\gamma : \mathbb{R} \longrightarrow C^\infty(\mathbb{R},\mathbb{R});\ \tilde{\gamma} \in C^\infty(\mathbb{R}^2,\mathbb{R})\} .$$

Hence $$\Phi = \{\varphi : C^\infty(\mathbb{R},\mathbb{R}) \longrightarrow \mathbb{R};\ \varphi\circ\gamma \in C^\infty(\mathbb{R},\mathbb{R}) \quad \forall\gamma \in \Gamma\} .$$

One knows (cf [15]) that $\Phi_{lin} = \{\varphi \in \Phi; \varphi \text{ linear}\}$ consists exactly of the distributions of compact support on $\mathbb{R}$. If one wants to verify (4), it is useful to show that one has even

$$(4') \quad \left.\begin{array}{l} \gamma : A \longrightarrow M \\ \varphi\circ\gamma \in M \quad \forall\varphi \in \Phi_{lin} \end{array}\right\} \Rightarrow \gamma \in \Gamma$$

The elegant proof due to Lawvere, Schanuel and Zame proceeds as follows: one first shows that the stronger property (4') holds for $M = \ell^{\infty}$, using the uniform boundedness principle. Using this one gets the result for $M = C^{\infty}(\mathbb{R},\mathbb{R})$ by means of the following lemma (cf. [7] and [13]).

Lemma. A function $f : \mathbb{R}^2 \longrightarrow \mathbb{R}$ is smooth iff all its partial difference quotients are $\underline{K}_{\ell^{\infty}}$-morphisms (i.e. are bounded on bounded sets).

Smooth spaces.

A smooth space is an object of the category $\underline{C}^{\infty}$ generated by $C^{\infty}(\mathbb{R},\mathbb{R})$. As an example of the consequences of the cartesian closedness of $\underline{C}^{\infty}$ we mention actions of smooth groups. A smooth group G is a smooth space with a compatible group structure (i.e. such that the group operations are smooth). For any smooth space X, the set $\mathrm{Diff}(X)$ of diffeomorphisms (i.e. $\underline{C}^{\infty}$-isomorphisms) of X becomes a smooth group if equipped with the smooth structure initial with respect to the two maps $i,j :: \mathrm{Diff}(X) \longrightarrow H(X,X)$ where $i(f) = f$ and $j(f) = f^{-1}$. An action of a smooth group G on a smooth space X is an action $G\pi X \longrightarrow X$ which is smooth. Cartesian closedness of $\underline{C}^{\infty}$ implies the

Proposition. There is bijection between the actions of G on X and the smooth homomorphisms $G \longrightarrow \mathrm{Diff}(X)$. This bijection is in fact a diffeomorphism with respect to the natural smooth structures of the respective function spaces.

If X is in particular a Banach manifold, one can show that $\mathrm{Diff}(X)$ becomes a subspace of $H(X,X)$. This means that oen can prove that the inversion map j is smooth with respect to the subspace structure by using an appropriate inverse function theorem.

For an arbitrary smooth space $X = (E,C,F)$ one can introduce tangent and cotangent spaces. Let $p \in E$ and $C_p = \{c \in C; c(0) = p\}$. One defines on C_p resp. F equivalence relations $\sim$ resp. $\sim_p$ as follows:

$$c_1 \sim c_2 \Leftrightarrow (f\circ c_1)^{\cdot}(0) = (f\circ c_2)^{\cdot}(0) \quad \forall f \in F ;$$

$$f_1 \sim_p f_2 \Leftrightarrow (f_1\circ c)^{\cdot}(0) = (f_2\circ c)^{\cdot}(0) \quad \forall c \in C_p .$$

$T_pX := C_p/\sim$ resp. $T^pX := F/\sim_p$ are called the tangent resp. cotangent space of X at p. From cartesian closedness it follows that F and hence also T^pX are in a natural way smooth vector spaces. T_pX can be imbedded in a (smooth) vector space (e.g. into the dual of T^pX, or into a vector space of derivations of F); but it is not always a vector space as the following example shows: X consists of two intersecting lines of $\mathbb{R}^2$ and carries the subspace structure, p is the intersection point.

One can consider various subcategories of $\underline{C}^\infty$, e.g. by imposing the condition that for each p, T_pX is a vector space, and that X is locally isomorphic to T_pX. Here, locally means with respect to the topology final with respect to the smooth curves. This topology has the good property that the condition for a map to be a $\underline{C}^\infty$-morphism is of local character and that inclusions of open subsets of an object belong to initial morphisms.

Calculus for convenient vector spaces

It is natural to develop calculus for vector spaces over the cartesian closed category $\underline{C}$, i.e. for smooth vector spaces. As in other cases (topological or compactly generated or pseudotopological vector spaces), some further restrictions have to be imposed in order to obtain theorems one wants to have. One restriction is in the direction of local convexity in order to have enough linear differentiable functions; the other is a completeness condition.

For any smooth vector space E we put

$$E' = \{\ell : E \longrightarrow \mathbb{R} \ ; \ \ell \text{ linear and smooth}\}$$

E' with its universal smooth structure is called the dual of E. The canonical map of E into the bidual is smooth.

<u>Definition</u>. A smooth vector space E is called <u>convenient</u> if

i) E' separates points;

ii) E' generates the smooth structure of E;

iii) E' yields a complete bornology on E.

Any Fréchet space with its natural smooth structure is convenient. In fact, its dual coincides with its topological dual, so i) holds. From theorem 1 follows ii). The bornology in iii) is formed by the subsets B of E for which $\ell(B)$ is bounded for all $\ell \in E'$. For a Fréchet space these subsets are the same as the bounded subsets of E in the usual sense. But for a locally convex space, bornological completeness (also called local completeness in [9]) is in general much weaker than usual completeness. This confirms a remark of Hogbe who said (cf. [8]) that nevertheless "for a great many problems bornological completeness turns out to be enough".

The conditions i), ii), iii) can be expressed equivalently by

i') The canonical map $E \to E''$ is injective;

ii') The smooth structure of E is initial with respect to $E \to E''$;

iii') The canonical map $E \to T_0E$ is surjective.

The map in iii') associates to $a \in E$ the tangent vector at 0 represented by the curve $\lambda \to \lambda \cdot a$.

Theorem 4. For a $\underline{C}$-morphism $f : E_1 \to E_2$ between convenient smooth vector spaces one has:

1) For all $a,h \in E_1$, $df(a,h) := \lim_{\lambda \to 0} \frac{f(a+\lambda h)-f(a)}{\lambda}$ exists;

2) $df(a,-)$ is linear;

3) $df : E_1 \pi E_1 \to E_2$ is also a $\underline{C}$-morphism and $E_1 \pi E_1$ is convenient;

4) The map $f \to df$ is a (linear) $\underline{C}$-morphism between the respective function spaces.

The limit in (1) is in the strong sense of Mackey-convergence; hence also for the weak topology induced on E_2 by E_2'. According to (3) one obtains higher order differentials $d^n f = d(d^{n-1}f)$ and one can prove for these the usual symmetry properties. Instead of working with df one can use $f' : E_1 \to L(E_1,E_2)$ defined as $f'(a)(h) = df(a,h)$ and $f^{(n)} = (f^{(n-1)})'$; $L(E_1,E_2)$ is again convenient. (4) implies that the smooth structure of $C^\infty(E_1,E_2)$ is initial with respect to the maps d^n, $n = 0,1,\ldots$

By showing that the conditions for convenient vector spaces are inherited by the function spaces $C^\infty(E_1,E_2)$ one obtains the first part of the following

Theorem 5. a) The category of convenient smooth vector spaces with the smooth maps as morphisms is cartesian closed.

b) The category of convenient smooth vector spaces with the linear smooth maps as morphisms has a tensor product making it a symmetric monoidal closed category.

In [11], A. Kriegl introduced certain locally convex spaces as the "right spaces for analysis in infinite dimension" and showed many interesting and useful properties of them; cf. also [12]. It is reassuring that these spaces can be identified with the convenient smooth vector spaces; cf. (1) and (4) in the following

Theorem 6. The following spaces can be identified with each other:

1) The convenient smooth vector spaces;
2) The vector spaces with compatible convenient ℓ^∞-structure;
3) The vector spaces with compatible convenient $C^{k,1}(\mathbb{R},\mathbb{R})$-structure;
4) The separated locally convex spaces which are bornological and bornologically complete;
5) The separated convex bornological vector spaces which are topological and bornologically complete;
6) The dual pairs (E,E') for which the bornology of E determined by E' is complete and for which E' satisfies one of the following equivalent conditions: the M-structure (C,F) of E generated by E' has the property $F_{Lin} = E'$ for
   a) $M = C^\infty(\mathbb{R},\mathbb{R})$,
   b) $M = C^{k,1}(\mathbb{R},\mathbb{R})$,
   c) $M = \ell^\infty$

In 2) and 3), convenient is defined as in the case of smooth vector spaces.

Using a general result of L. Nel [14] one can show that for any smooth space X the space F_X of smooth functions $X \to \mathbb{R}$ is the dual of a smooth vector space L_X ; in fact $X \to L_X$ belongs to a coadjoint to the forgetful functor from smooth vector spaces to smooth spaces. The dual of any smooth vector space being convenient, F_X is always convenient. Moreover, F_X being a dual F_X'' is a tri-dual and one therefore gets a retraction to the canonical map $F_X \to F_X''$. Hence F_X is reflexive iff L_X is Mackey-dense in F_X' . For X a classical manifold, F_X' consists of the distributions of compact support on X and L_X of the elementary distributions (i.e. linear combinations of evaluations at points), and in this case A. Kriegl could show the respective density and hence the reflexivity of F_X (seminar on smooth functions, University of Geneva, not published).

R E F E R E N C E S

[1] A. Bastiani: "Applications différentiables et variétès différentiables de dimension infinie". Journal d'Analyse mathématique XIII, p. 1-114.

[2] J. Boman: "Differentiability of a function and of its compositions with functions of one variable", Math. Scand. 20, 1967, p. 249-268.

[3] J. F. Colombeau: "Différentiation et Bornologie". Thèse, Université de Bordeaux I, 1973.

[4] A. Frölicher and W. Bucher: "Calculus in Vector Spaces without Norm", Lecture Notes in Math. 30, Springer 1966.

[5] A. Frölicher: "Catégories cartésiennement fermées engendrées par des monoïdes", Cahiers de Top. et Géom. diff. XXI/4, 1980, p. 367-375.

[6] A. Frölicher: "Applications lisses entre spaces et variétés de Fréchet" C. R. Ac. Sci. Paris 293, 1981, p. 125-127.

[7] Haupt, Aumann, Pauc: "Differential- und Integralrechnung Bd II, Göschens Lehrbücherei Band 25, de Gruyter 1950.

[8] H. Hogbe-Nlend: "Bornologies and functional analysis", Mathematics Studies 26, North-Holland 1977.

[9] H. Jarchow: "Locally convex spaces", Teubner 1981.

[10] A. Kriegl: "Eine Theorie glatter Mannigfaltigkeiten und Vektorbünder; Dissertation, Wien 1980.

[11] A. Kriegl: "Die richtigen Räume für Analysis im unendlichdimensionalen", to appear in Monatshefte für Mathematik.

[12] A. Kriegl: "Eine kartesisch abgeschlossene Kategorie glatter Abbildungen zwischen beliebigen lokalkonvexen Vektorräumen", preprint 1982.

[13] F. W. Lawvere, S. H. Schanuel and W. R. Zame: "On C^∞ Function Spaces", Preprint 1981.

[14] L. Nel: "Convenient topological Algebra and reflexive objects", in Lecture Notes in Math. Vol. 719, p. 259 ff.

[15] N. van Que and G. Reyes: "Théorie des distributions et théorèmes d'extension de Whitney", Exposé 8, Géom. diff. synth. fasc. 2, Rapport de Recherches DMS 80-12, Universite de Montréal 1980.

[16] U. Seip: "A convenient Setting for Smooth Manifolds". J. of pure and appl. Algebra 21, 1981, p. 279-305.

INTRODUCTION TO SYNTHETIC DIFFERENTIAL GEOMETRY, AND A SYNTHETIC THEORY OF DISLOCATIONS

Anders Kock
Department of Mathematics
Aarhus University
Aarhus, Denmark

This article is divided into two parts, corresponding to the two lectures I gave at the conference. The first expounds some of the category theoretic approach to differential geometry (Lawvere, Kock, Wraith, Reyes, Dubuc, Joyal et al.). A fuller account may be found in [7], but we want to emphasize that a fair amount of geometric theory can be developed just on the basis of what we shall now present. This in particular applies to the second part, concerning dislocations in crystals. This theory I learned by reading Noll's [12], and it was presented first in synthetic form in my [8].

LECTURE 1: INTRODUCTION TO SYNTHETIC DIFFERENTIAL GEOMETRY

We want to have an axiomatic mathematical theory of continua, or of smooth objects, by giving a theory of the totality of such.

So we would not start by saying "a smooth object is a set of elements (atoms) with some additional structure"; but rather, smoothness is a property of how these objects relate to each other, or map to each other.

This forces us to have the notion of map as the primitive concept, and how maps compose - in other words, to have a category $\mathcal{E}$ as basic datum.

In a category $\mathcal{E}$, an object $M \in \mathcal{E}$ is not determined by its 'atoms', meaning maps in $\mathcal{E}$ of form $\mathbb{1} \longrightarrow M$ (where $\mathbb{1}$ is 'the' one-element object, i.e. 'the' terminal object of $\mathcal{E}$, see below), just as a continuous body is not determined by its collection of atoms. However, M is determined by the totality of all maps in $\mathcal{E}$ mapping into it

$$b: X \longrightarrow M ,$$

X arbitrary in $\mathcal{E}$. Example: if X is I, an interval on the line (to be introduced axiomatically shortly), b is a curve in M. If b is injective, it may be interpreted as a sub-body, and if X is sufficiently small, as a small but extended body-element of M . Generally, we may think of b as an X-parametrized family of 'elements' of M , but the parameter domain is not a discrete set.

So we adopt the terminology that any $b: X \longrightarrow M$ is a

$$\left.\begin{array}{l}\text{parametrized element of } M \\ \text{or} \\ \text{generalized element of } M\end{array}\right\} \text{ with domain } X$$

(or 'defined at stage X').

One advantage of this terminology is that now the notion, say, of cartesian product $M \times N$ of two objects M and N can be defined (up to isomorphism) by saying: $M \times N$ is an object whose (generalized) elements (at any stage X) are in 1-1 correspondence with pairs of elements, from M and N , respectively (likewise at stage X)

$$\frac{X \longrightarrow M \times N}{X \longrightarrow M \qquad X \longrightarrow N}\,.$$

Besides products, the category $\mathcal{E}$ should allow existence of function-space objects, e.g. path spaces M^I ; generally, for any pair A, B of objects of $\mathcal{E}$, there should exist a function-space object B^A , whose elements at stage X are in 1-1 correspondence with maps $X \times A \longrightarrow B$ (= X-parametrized families of maps $A \longrightarrow B$) , i.e., there should be a natural bijective correspondence

$$\frac{X \longrightarrow B^A}{X \times A \longrightarrow B}$$

("exponential adjointness", or "λ-conversion").

Also, $\mathcal{E}$ should have a terminal object $\mathbb{1}$ - this is an object with exactly one element at each stage: $X \;\; ! \; X \longrightarrow \mathbb{1}$. With these assumptions, $\mathcal{E}$ is what is called a Cartesian Closed Category.

We assume that there is given in $\mathcal{E}$ a commutative ring object R , to be thought of as the ring "of pure quantities", or, via some standard non-canonical identifications, as the geometric line, with a "zero" 0 and a "one" 1 chosen on it.

To say that R is a commutative ring object means that there are given addition and multiplication laws which allow us to add and multiply (generalized) elements defined at the same parameter domain: to

$$X \xrightarrow{a} R \qquad X \xrightarrow{b} R$$

associate

$$X \xrightarrow{a+b} R$$

(and similarly for multiplication), with a naturality condition. Equivalently, by a standard theorem of category theory, there is given an addition map

$$R \times R \xrightarrow{+} R \ ,$$

as well as a multiplication map $R \times R \xrightarrow{\cdot} R$, and also $\mathbb{1} \xrightarrow{\ulcorner 0 \urcorner} R$ and $\mathbb{1} \xrightarrow{\ulcorner 1 \urcorner} R$ (zero- and one-elements), with suitable associative, distributive etc. laws.

In the models $\mathcal{E}$,R for the axiomatics (cf. e.g. the lecture by Reyes [13]), $\mathcal{E}$ will contain the category of smooth (= C^{∞}-) manifolds as a subcategory, and R will be $\mathbb{R}$, but it gets different properties when viewed in $\mathcal{E}$; for instance, it will acquire (generalized!) non-zero nilpotent elements r ($r^n = 0$) .

Returning to the abstract situation, it makes sense to say that an element

$$X \xrightarrow{d} R$$

has square zero, $d^2 = 0$; diagrammatically

$$\begin{array}{ccc} X & \xrightarrow{(d,d)} & R \times R \\ \downarrow & & \downarrow \cdot \\ \mathbb{1} & \xrightarrow[0]{} & R \end{array}$$

commutes.

We shall assume that $\mathcal{E}$ has <u>equalizers</u>; for the moment, it suffices even to ask that there is an injective (= monic) map

$$D \hookrightarrow R$$

which "classifies" (or "consists of") those elements of R which have square zero, $D = [\![x \in R \mid x^2 = 0]\!]$, or, diagrammatically: for any X

$X \xrightarrow{d} R$ factors through $D \hookrightarrow R$, i.e.

$$\begin{array}{ccc} D & \hookrightarrow & R \\ \uparrow & \nearrow_{d} & \\ X & & \end{array}$$

iff

d has square zero.

(Saying that $\mathcal{E}$ has equalizers amounts essentially to saying that we can form $[\![x \in A \ldots]\!]$ where "..." stands for any <u>equation</u> in x . We use the notation $[\![x \in R | x^2 = 0]\!]$ rather than $\{x \in R | x^2 = 0\}$ to emphasize that this is not a set, but an object of $\mathcal{E}$. However it is an experience in category theory, which is also supported by a <u>systematic semantics</u> (see e.g. [9] or [7]) that one can work with such objects "as if" they were the sets which the notation suggests. We can for instance define a map

$$D \times R \xrightarrow{\cdot} D$$

by giving the following <u>description</u> of it, and argument for why it lands in D:

$$(d,b) \longmapsto d \cdot b \text{ ; belongs to } D \text{ ,}$$

$$\text{because } (d \cdot b)^2 = d^2 \cdot b^2 = 0 \cdot b^2 = 0 \text{ .}$$

(Of course, it would in this case not be difficult diagrammatically to prove that we have a factorization

$$\begin{array}{ccc} R \times R & \xrightarrow{\quad\cdot\quad} & R \\ \uparrow & & \uparrow \\ D \times R & \dashrightarrow & D \end{array} \text{ ,}$$

the lower map of which being the one given by the above description.)

This systematic use (or misuse) of set-theoretic notation will be used without further comments.

For instance, we can construct a map, crucial in the statement of Axiom 1 below,

$$R \times R \times D \longrightarrow R$$

$$(a,b,d) \longmapsto a + d \cdot b .$$

Its exponential adjoint is denoted α :

$$R \times R \xrightarrow{\alpha} R^D \text{ ;}$$

it may be given by the formula

$$(a,b) \longmapsto [d \longmapsto a + d \cdot b] \text{ .}$$

We can now state the main Axiom, which will carry a good deal of differential calculus on its shoulders:

<u>Axiom 1</u> α is invertible ("bijective").

Allowing unhindered use, or misuse, of set theoretical language and English language (justifiable, as argued above), the axiom can be read:

"Every function $f: D \longrightarrow R$ is of form
$f(d) = a + d\cdot b \quad \forall d \in D$, for unique a and b in R" §

or

"Every function $f: D \longrightarrow R$ extends to a unique
affine (=flat) function $R \longrightarrow R$" §§

Alternatively, we may formulate the axiom in geometric form:

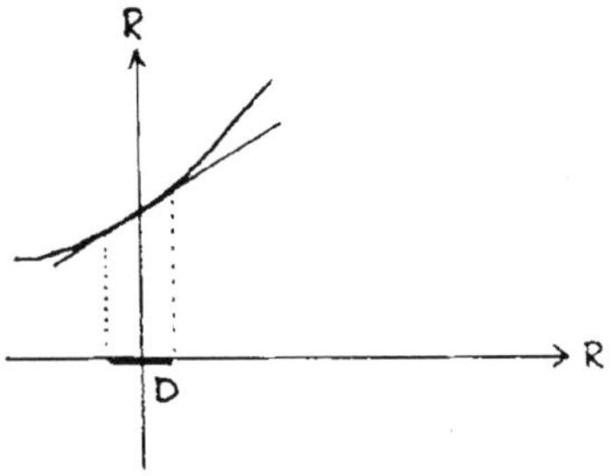

§§§

"D is small enough so that any function defined on D is affine, but <u>large enough</u> to determine the <u>slope</u> uniquely".

The <u>uniqueness</u> assertion for b (the slope) is very strong; we may isolate it as follows:

$$(\forall d \in D: d\cdot b_1 = d\cdot b_2) \Rightarrow (b_1 = b_2) ,$$

which we express verbally by saying

"universally quantified d's may be cancelled".

Since the notion of <u>slope</u> is thus built in, we can, for any $f: R \longrightarrow R$, define its <u>derivative</u> $f': R \longrightarrow R$; we define, for $x \in R$, $f'(x)$ as the unique element of R for which

$$f(x+d) = f(x) + d\cdot f'(x) \qquad \forall d \in D$$

("Taylor's formula").

From Taylor's formula in this form, a fair amount of differential calculus follows purely algebraically (chain rule, Leibniz rule,...); see e.g. [7].

At this point we should warn the reader that the usage of set-theoretic reasoning in a general cartesian closed category like $\mathcal{E}$ has limitations. In particular, this applies to the logical principle called the law of excluded middle; this principle is at work in a construction of the Dirac delta function ("value 1 if $x = 0$, value 0 if $x \neq 0$"), and the existence of such functions is quite quickly seen to be incompatible with Axiom 1. Alternatively, Axiom 1 is so strong as to be inconsistent when classical logic is used. It is possible and not difficult to describe the weaker logic that applies in all cartesian closed categories (roughly, it is the so-called intuitionistic logic). With this weaker logic, Axiom 1 is no longer inconsistent.

The moral to be learned here, and which we believe is new in mathematics, is that weaker logics have their virtue, not by being more "secure" or "trustworthy" than full classical logic, but by their ability to carry theories (like Axiom 1) which are so strong as to be inconsistent in classical logic.

We return to the development of some aspects of differential geometry in $\mathcal{E}$, with an R that satisfies Axiom 1.

It turns out to be correct to define a tangent vector to an object $M \in \mathcal{E}$ as an arbitrary map

$$t: D \longrightarrow M .$$

The element $t(0) \in M$ is called the base point of the tangent vector t. The function space object M^D is thus the (total space of) the tangent bundle of M, and the map $M^D \longrightarrow M$ given by $t \longmapsto t(0)$ makes it a bundle over M, which we also, following classical notation, denote $TM \longrightarrow M$. The fibre of this over an $x \in M$, i.e. the set of all $t: D \longrightarrow M$ with $t(0) = x$, is denoted $T_x M$.

One indication why this definition of tangent bundle is correct, is that Axiom 1 now in particular expresses that R has the correct tangent bundle: $TR = R^D \cong R \times R$. Similarly, it follows that $T(R^n) \cong R^{2n}$, and that in fact $T_{\underline{x}}(R^n) \cong R^n$ for any $\underline{x} \in R^n$. (For general M, a few assumptions on it will provide each $T_x M$ with a canonical R-module structure.)

We now present some strengthenings of Axiom 1, related to some higher-dimensional and higher-order infinitesimal objects. Consider namely $D(n) \hookrightarrow R^n$ given by

$$D(n) := [\![(x_1,\ldots,x_n) \in R^n \mid x_i \cdot x_j = 0 \quad \forall i,j = 1,\ldots,n]\!] ,$$

and more generally $D_k(n) \hookrightarrow R^n$ (for $k \geq 1$ an integer)

$$D_k(n) := [\![(x_1,\ldots,x_n) \in R^n \mid \text{the product of any } k+1 \text{ of the } x_i\text{'s is } 0]\!] .$$

The strengthened Axiom 1 says, in verbal form,

<u>Axiom 1'</u> Any function $D_k(n) \longrightarrow R$ extends uniquely to a polynomial function in n variables $R^n \longrightarrow R$ of degree $\leq k$.

For $n = 1$ and $k = 1$, $D_k(n) = D$, and we recover the old Axiom 1. For n arbitrary, but $k = 1$, we have in particular that any $D(n) = D_1(n) \longrightarrow R$ extends uniquely to an affine (= flat) map $R^n \longrightarrow R$.

It is trivial to derive that similar extension properties hold for maps $D_k(n) \longrightarrow R^m$ for arbitrary m .

For $\underline{x} \in R^n$, we denote by $\mathfrak{m}_k(\underline{x}) \hookrightarrow R^n$ the subset

$$\mathfrak{m}_k(\underline{x}) := [\![\underline{y} \in R^n \mid \underline{y} - \underline{x} \in D_k(n)]\!] ,$$

"the k-<u>monad</u> around $\underline{x}$" . (It represents, in a definite way, the notion of k-<u>jet</u> at $\underline{x}$.) Call a subset $U \hookrightarrow R^n$ <u>formally open</u> if for all k, $\underline{x} \in U$ implies $\mathfrak{m}_k(\underline{x}) \hookrightarrow U$. (Such U are examples of <u>formal</u> n-<u>dimensional</u> <u>manifolds</u>.) They carry a binary relation on them given by

$$\underline{x} \sim \underline{y} \quad \text{iff} \quad \underline{y} \in \mathfrak{m}_1(\underline{x}) \quad \text{iff} \quad \underline{x} - \underline{y} \in D(n) ,$$

called the <u>neighbour</u>-relation. It is clearly reflexive and symmetric, but not transitive: note that $d_1^2 = d_2^2 = 0$ does not imply $d_1 \cdot d_2 = 0$, hence does not imply $(d_1 + d_2)^2 = 0$. However, $(d_1 + d_2)^3 = 0$, clearly.

More generally, the addition map $R^n \times R^n \xrightarrow{+} R^n$ restricts to a map

$$D(n) \times D(n) \xrightarrow{+} D_2(n) ,$$

which will be of crucial importance in the next lecture. We have for it the following

<u>Proposition 1.1</u> For any map $\tau : D(n) \times D(n) \longrightarrow R$ which is

symmetric $(\tau(\underline{d},\underline{\delta}) = \tau(\underline{\delta},\underline{d}))$, there exists a unique map $t\colon D_2(n) \longrightarrow R$ such that $\tau(\underline{d},\underline{\delta}) = t(\underline{d} + \underline{\delta})$ $\forall(\underline{d},\underline{\delta}) \in D(n) \times D(n)$.

Proof. We do the case $n = 1$ only. From Axiom 1 easily follows that any $\tau\colon D \times D \longrightarrow R$ is of form

$$\tau(d_1,d_2) = a + d_1\cdot b_1 + d_2\cdot b_1 + d_1\cdot d_2\cdot c .$$

If τ is assumed symmetric, we get $\tau(d,0) = \tau(0,d)$, $\forall d \in D$, so

$$a + d\cdot b_1 = a + a\cdot b_2 \qquad \forall d \in D .$$

Cancelling the universally quantified d, we get $b_1 = b_2$ $(= b$, say$)$, so

$$\begin{aligned}\tau(d_1,d_2) &= a + (d_1 + d_2)\cdot b + d_1\cdot d_2\cdot c\\ &= a + (d_1 + d_2)\cdot b + \tfrac{1}{2}(d_1 + d_2)^2\cdot c\\ &= t(d_1 + d_2) ,\end{aligned}$$

where $t(\delta) = a + \delta\cdot b + \frac{1}{2}\delta^2\cdot c$ $\quad\forall\delta \in D_2$. This proves the existence of t. Assume t' also satisfies the conclusion. By Axiom 1' (for $n = 1$, $k = 2$), t' is given by $t'(\delta) = a' + \delta\cdot b' + \frac{1}{2}\delta^2\cdot c'$. By assumption, we get

$$a + (d_1 + d_2)\cdot b + d_1\cdot d_2\cdot c = a' + (d_1 + d_2)\cdot b' + d_1\cdot d_2\cdot c'$$

$\forall(d_1,d_2) \in D \times D$. For $d_1 = d_2 = 0$, we get $a = a'$. For $d_2 = 0$, we get $d_1\cdot b = d_1\cdot b'$ $\forall d_1 \in D$, whence $b = b'$, by cancelling the universally quantified d_1. So $d_1\cdot d_2\cdot c = d_1\cdot d_2\cdot c'$ $\forall d_1,d_2$, whence cancelling d_1 and d_2 one at a time yields $c = c'$.

Let us say that an object M has the <u>symmetric functions property</u> if any symmetric $\tau\colon D(n) \times D(n) \longrightarrow M$ factors uniquely across $D(n) \times D(n) \xrightarrow{+} D_2(n)$; then the Proposition expresses that R has the symmetric functions property. It is immediate to conclude, then, that so does R^k, as well as any formally open $U \hookrightarrow R^k$. (In fact, any formal manifold, and any 'affine scheme' will have this property.)

We utilize Axiom 1' (for $k = 1$) to give a non-classical notion of differential 1-form on formal manifolds - for simplicity of exposition, we do it only for formally open $M \hookrightarrow R^n$.

Definition An R^k-valued 1-form ω on M is a law, which to each pair of neighbours $x \sim y$ in M associates an element $\omega(x,y) \in R^k$, and such that $\omega(x,x) = 0$ $\forall x \in M$.

If $g: M \longrightarrow R^k$ is a function, we define a 1-form dg by

$$(dg)(x,y) = g(y) - g(x) .$$

Such 1-forms are called exact.

If ω is a 1-form on M, we say that it is closed if $d\omega = \underline{0}$ meaning: for any triple x,y,z of mutual neighbours in M,

$$\omega(x,y) + \omega(y,z) + \omega(z,x) = 0 .$$

We say that a 1-form is alternating if $\omega(x,y) = -\omega(y,x)$ for all $x \sim y$; for many objects, for instance for formally open $M \hookrightarrow R^n$, this can quite easily be proved to be a consequence of $\omega(x,x) = 0 \ \forall x$.

The fact that this 1-form notion agrees with the classical one (formulated synthetically), and likewise that the use of the word 'frame' (in the next lecture) is justified, follows from the following, closely related, propositions.

Proposition 1.2 Given $x \in M$. There is a natural bijective correspondence between

$$\frac{\text{maps } \mathfrak{m}(x) \xrightarrow{f} R \text{, taking } x \text{ to } 0}{R\text{-linear maps } T_xM \xrightarrow{F} R \text{ .}}$$

and

(We write $\mathfrak{m}(x)$ for $\mathfrak{m}_1(x)$.)

Proposition 1.3 Given $x \in M$. There is a natural bijective correspondence between

$$\frac{\text{maps } D(n) \xrightarrow{k} M \text{, taking } \underline{0} \text{ to } x}{R\text{-linear maps } R^n \xrightarrow{K} T_xM} ;$$

and

any such k maps $D(n)$ into $\mathfrak{m}(x)$, and K is bijective iff $k: D(n) \longrightarrow \mathfrak{m}(x)$ is bijective.

We only indicate the correspondence, and the idea of proof, for the first of the propositions. Given f, we want to construct $F: T_xM \longrightarrow R$, so let $t \in T_xM$, i.e. $t: D \longrightarrow M$ with $t(0) = x$. We define $F(t) \in R$ by

$$d\cdot F(t) := f(t(d)) \quad \forall d \in D ;$$

this makes sense, since the right hand side, as a function of $d \in D$

takes 0 to 0, and hence, by Axiom 1, is of form $d \longmapsto d \cdot b$ for a unique b, and this b we denote $F(t)$. Note that the passage from f to F does not involve coordinates. Going backwards from F to f, however, does. Roughly, the steps in the bijection are

$$\frac{\dfrac{\dfrac{T_xM \longrightarrow R \qquad \text{R-linear}}{R^n \longrightarrow R \qquad \text{R-linear}} \; (\text{by } T_xM \cong R^n)}{D(n) \longrightarrow R \qquad \underline{0} \longmapsto 0} \; \text{by Axiom 1'}}{\mathfrak{m}(x) \longrightarrow R \qquad x \longmapsto 0\,.} \; (\text{by } D(n) \cong \mathfrak{m}(x))$$

Let us remark that: giving an $f\colon \mathfrak{m}(x) \longrightarrow R$ with $f(x) = 0$ for each x is of course equivalent to giving $\omega(x,y) = f(x)(y)$ for each neighbour pair $x \sim y$, and with $\omega(x,x) = 0$, thus is equivalent to giving a 1-form in our sense; whereas giving an R-linear $T_xM \longrightarrow R$ for each x amounts to the classical definition of 1-form. The same applies (with suitable generalization of Proposition 1.2) to 1-forms with values in R^k, say. (For more about the differential form notion defined in terms of the neighbour relation, we refer to unpublished work of Bkouche and Joyal, and to [5], [6].)

LECTURE 2: A SYNTHETIC THEORY OF DISLOCATIONS

In the following, n is a fixed integer, typically $n = 2$ or 3 for the intended application. We shall deal with an object (or 'body') M which shares with R^n and formally open subsets thereof the structure of having a reflexive and symmetric relation $\sim$ defined on it, called the <u>neighbour</u> relation. For $x \in M$, we denote by $\mathfrak{m}(x) \hookrightarrow M$ the subset

$$\mathfrak{m}(x) := [\![y \in M \mid x \sim y]\!]\ .$$

We call it the <u>monad</u> (or 1-monad) <u>around</u> x. (It is in fact possible, cf [4], to define canonically on any object $M \in \mathcal{E}$ such a relation. So it is really not an added structure.) For R^n, and the relation $\sim$ defined in the previous lecture ($\underline{x} \sim \underline{y}$ iff $\underline{x}-\underline{y} \in D(n)$) each $\mathfrak{m}(x)$ is isomorphic to $D(n)$ via the translation map $D(n) \longrightarrow \mathfrak{m}(x)$ given by $\underline{d} \longmapsto \underline{x}+\underline{d}$. This map is a <u>frame</u> in the following more general sense: For the general object M under consideration, we call a map

$$h : D(n) \longrightarrow \mathfrak{M}(x)$$

a frame at x if it is bijective and takes $\underline{0}$ to x. For technical reasons, we shall further assume that, for $\underline{d}$ and $\underline{\delta} \in D(n)$, we have

$$h(\underline{d}) \sim h(\underline{\delta}) \quad \text{iff} \quad \underline{d}-\underline{\delta} \in D(n) \;; \tag{2.1}$$

for many M in the models for the theory, this will be automatic. (The inverse h^{-1} of a frame at x is to be thought of as a configuration gradient at x, or a local configuration at x, to use terminology of [12]. This follows from Proposition 1.3.)

By a framing k of M, we mean a law which associates to each x a frame $k_x : D(n) \longrightarrow \mathfrak{M}(x)$ at x. (The collection of the k_x^{-1} would in the terminology of [12] be a reference for M.) We don't have to say explicitly that the law is smooth, because the whole thing takes place in $\mathcal{E}$, a category "of smooth sets"; the framing itself being a map in $\mathcal{E}$ from M to $M^{D(n)}$.

Considerations of references, or framings, on objects like M are motivated by the consideration of crystals: in a crystal, the lattice structure of molecules in it will, even when the crystal is not perfect, determine a frame ("infinitesimal coordinate system") at each point; precisely the imperfections or dislocations prevent, or obstruct, existence of a global coordinate system.

(Contemplation of a picture that appears in [11] will convey some idea of how dislocations might look microscopically (for $n = 2$). Or take any book on engineering materials: dislocations are crucial for the plastic deformation of steel.)

Of course, the differential-geometric theory of [1], [11], [12], represent a macroscopic viewpoint. This is also the case for the theory I present here: even though it talks about "neighbour points", these are not to be interpreted "microscopically", i.e. as neighbour molecules. Our synthetic theory is still macroscopic: it does not put the blame (or credit) for the dislocations on a few individual molecules, but considers the dislocations as being continuously distributed over the body M.

To say that the crystal is perfect would say that the framing arising from the crystal structure comes about from an (almost-) global configuration, more precisely, from an immersion g as in the following definition:

Definition. A framing k of M is called homogeneous if there exists a map ("immersion") $g : M \longrightarrow R^n$ such that for any $x \in M$, g restricted to the monad around x is inverse to k_x modulo translation by $g(x) \in R^n$, or, equivalently, if for any x the diagram

$$\begin{array}{ccc} \mathfrak{m}(x) & \xrightarrow{k_x^{-1}} & D(n) \\ \downarrow & & \downarrow +g(x) \\ M & \xrightarrow[g]{} & R^n \end{array} \qquad (2.2)$$

commutes.

We want to give infinitesimal criteria for when a framing k is homogeneous.

We first state our assumptions concerning M. They are weak, for instance in the models [2], [3], [7], [13], they will be satisfied by any smooth manifold, and in Lecture 1, they were shown to hold for R^m on basis of certain versions of Axiom 1.

Assumptions:

(i) M has the symmetric functions property

(ii) R^n-valued 1-forms on M are alternating.

(A third, stronger assumption, essentially that M is simply connected, will be made for the final result below.)

To a given framing k of M, we now associate an R^n-valued 1-form $\omega = \omega_k$ on M:

$$\text{for } x \sim y,\ \omega(x,y) := k_x^{-1}(y) \in D(n) \hookrightarrow R^n$$

(read: "$\omega(x,y)$ is the coordinates of y in the infinitesimal coordinate system k_x around x").

<u>Proposition 2.1</u> If ω_k is exact, then k is homogeneous.

<u>Proof</u>. By assumption, $\omega = dg$ for some $g : M \longrightarrow R^n$. So we have, for $x \sim y$,

$$k_x^{-1}(y) = \omega(x,y) = (dg)(x,y) = g(y) - g(x) \ ;$$

add $g(x)$ to both sides of this equation to get commutativity of diagram (2.2).

We next consider the "infinitesimal parallel transport" $\lambda = \lambda_k$ to which the "crystal structure", i.e. the framing k, gives rise. Precisely, we introduce a ternary operation λ on M, $\lambda(x,y,z)$ being defined whenever $x \sim y$ and $x \sim z$:

$$\lambda(x,y,z) := k_y(k_x^{-1}(z)) ,$$

(read: "the point which in the y-coordinate system has the same coordinates as z has in the x-coordinate system"). A figure will do no harm:

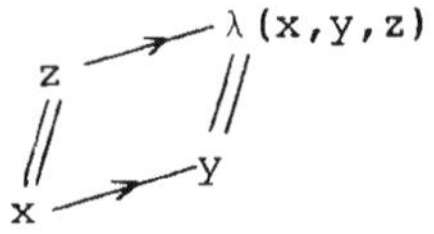

(2.3)

It indicates that, geometrically, $\lambda(x,y,z)$ closes a parallellogram with sides xy and xz, but it need not be symmetric in y, z, which is why we use different signatures for the two pairs of lines in the figure. (Note that we do not, for the construction of λ, have to insist on $y \sim x$, which we do only to conform with the general connection notion of [8].)

We say λ is <u>torsion free</u>, or <u>symmetric</u>, if

$$\lambda(x,y,z) = \lambda(x,z,y) \quad \forall x, \forall y \sim x, \forall z \sim x .$$

<u>Proposition 2.2</u> If λ_k is torsion free, then ω_k is closed.

<u>Proof</u>. Consider a triangle x,y,z of mutual neighbours in M,

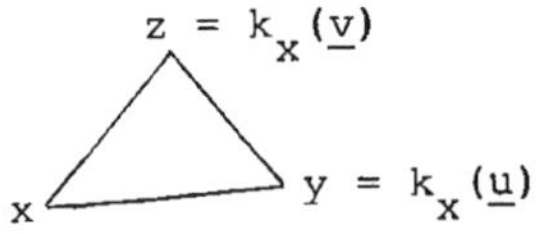

with $\underline{u}$ and $\underline{v} \in D(n)$. We must prove $(d\omega)(x,y,z) = \underline{0}$. Since $z \sim y$ the assumption (2.1) applied to k_x yields $\underline{v}-\underline{u} \in D(n)$. Consider the map

$$D(n) \times D(n) \longrightarrow M$$

$$(\underline{d},\underline{\delta}) \longmapsto k_{k_x(\underline{d})}(\underline{\delta}) = \lambda(x,k_x(\underline{d}),k_x(\underline{\delta})) .$$

By assumption on λ, this is symmetric in $(\underline{d},\underline{\delta})$, hence, by the Symmetric Functions Property for M, there is a factorization

$$\begin{array}{ccc} D(n) \times D(n) & \longrightarrow & M \\ {\scriptstyle +}\downarrow & \nearrow_{k'} & \\ D_2(n) & & \end{array} \quad .$$

So

$$k_{k_x(\underline{d})}(\underline{\delta}) = k'(\underline{d} + \underline{\delta}) \qquad \forall(\underline{d},\underline{\delta}) \in D(n) \times D(n) \tag{2.4}$$

Note that $\forall \underline{\delta} \in D(n)$

$$k'(\underline{\delta}) = k_x(\underline{\delta}) \tag{2.5}$$

(put $\underline{d} = \underline{0}$ in (2.4)).

We now calculate:

$$(d\omega)(x,y,z) = \omega(x,y) + \omega(y,z) + \omega(z,x) =$$

$$k_x^{-1}(y) + k_y^{-1}(z) + (-k_x^{-1}(z))$$

(because $\omega(z,x) = -\omega(x,z)$ by assumption (ii)). The two outer terms here are $\underline{u}$ and $-\underline{v}$, respectively, so to prove $d\omega(x,y,z) = \underline{0}$, we just need to show that

$$k_y^{-1}(z) = \underline{v} - \underline{u} \ ,$$

or equivalently

$$z = k_y(\underline{v} - \underline{u}) \tag{2.6}$$

(the right hand side makes sense, since $\underline{v} - \underline{u} \in D(n)$, as observed). Now, by (2.4),

$$k_y(\underline{v} - \underline{u}) = k_{k_x(\underline{u})}(\underline{v} - \underline{u}) = k'(\underline{u} + (\underline{v} - \underline{u})) \ ,$$

but

$$k'(\underline{u} + (\underline{v} - \underline{u})) = k'(\underline{v}) = k_x(\underline{v}) = z$$

by (2.5), proving (2.6) and thus the proposition.

We now derive the main result by combining these two propositions with one further assumption, which essentially is an inegration axiom [6], [10] and a simply-connectedness assumption on M, [10]:

(iii) closed R^n-valued 1-forms on M are exact.

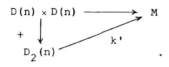

We then have

Theorem 2.3 If a framing k has the associated connection λ_k torsion free, it is homogeneous.

Proof. If λ_k is torsion free, ω_k is closed by Proposition 2.2, hence exact, hence k is homogeneous by Proposition 2.1.

Remark. The converse implications in the proof trivially hold. Also, with suitable assumptions, one may prove the essential uniqueness of an immersion g witnessing homogeneity of k.

The interest of this theorem in dislocation theory for crystals is that the torsion of λ_k can be geometrically seen, and drawn, in terms of the so-called "Burgers vector of a circuit", which has been explained in terms of the sum of all small vectors making up a circuit, this sum being zero iff λ is torsion free. In our context, for x,y,z as in (2.3), we want to investigate to what extent

$$\lambda(\lambda(x,y,z),y,z) = x . \tag{2.7}$$

We may view the vector $\overline{x\,\lambda(x,y,z)}$ as the "sum" of $\overline{xy}$ and $\overline{xz}$ according to λ; then (2.7) is one way of giving meaning to the phrase "sum of the four sides of the parallelogram (2.3) being zero", and the failure of (2.7) as the Burgers vector. However, (2.7) holds if λ is torsion free, by the following calculation. First we prove that for any $\lambda = \lambda_k$ arising from a framing k, we have

$$\lambda(\lambda(x,y,z),z,y) = x . \tag{2.8}$$

Let $w = \lambda(x,y,z)$, so $w = k_y(k_x^{-1}(z))$, so

$$k_y^{-1}(w) = k_x^{-1}(z) ,$$

whence, since ω_k is alternating

$$k_w^{-1}(y) = k_z^{-1}(x) ,$$

or $k_z(k_w^{-1}(y)) = x$, which is (2.8). If λ is torsion free, (2.8) implies (2.7).

Besides for crystals, dislocations occur in geography, and are for instance visible from an airplane when flying across the North American prairie, as I had the privilege of doing in 1972; this was over Alberta, where the effect is more pronounced due to the latitude. My co-passenger pointed out to me 1) how the prairie was divided by roads into squares 1 mile by one mile; and 2) how at regular intervals, some of the roads running North-South made what he called an <u>off-set</u> of up to hundred yards:

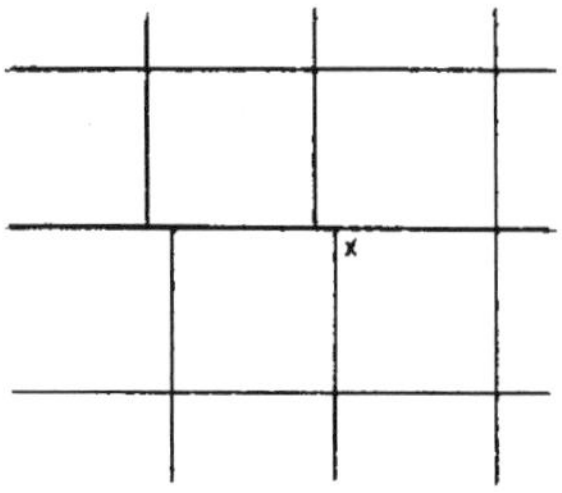

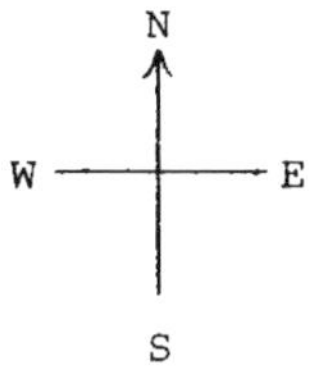

This off-set is actually a Burgers vector witnessing the inhomogeneity of a certain framing k, i.e. the torsion of the associated connection λ_k. The framing k is the "yardstick-and-compass" - framing: at each point of the prairie, the compass tells you where North and East are, and with the yardstick, you then have a local coordinate system, or a frame.

If one walks from x one mile East, then one mile North, then one mile West, and finally one mile South, one will end up somewhat to the West of x, and this "somewhat" is the off-set, Burgers vector or failure of (2.7).

REFERENCES

1. B. A. Bilby, Continuous distribution of dislocations, Progr. in solid mechanics 1 (1959).

2. E. J. Dubuc, Sur les modèles de la géometrie differentielle synthétique, Cahiers de top. et géom. diff. 20 (1979), 231-279.

3. E. J. Dubuc, C^∞-schemes, American Journ. Math. 103 (1981), 683-690.

4. E. J. Dubuc and A. Kock, On 1-form classifiers, Aarhus Preprint Series 1981/82 No. 39.

5. A. Kock, Differential forms with values in groups (preliminary report), Cahiers de top. et géom diff. 22 (1981), 141-148.

6. A. Kock, Differential forms with values in groups, Bull. Austr. Math. Soc. 25 (1982), 357-386.

7. A. Kock, Synthetic differential geometry, London Math. Soc. Lecture Notes Series, 51, Cambridge University Press 1981.

8. A. Kock, A combinatorial theory of connections, Aarhus Preprint Series 1981/82 No. 24.

9. A. Kock and G. E. Reyes, Doctrines in categorical logic, in Handbook of mathematical logic (ed. J. Barwise), North Holland Publishing Company 1977.

10. A. Kock and G. E. Reyes, Models for synthetic integration theory, Math. Scand. 48 (1981), 145-152.

11. E. Kröner, Allgemeine Kontinuumstheorie der Versetzungen und Eigenspannungen, Arch. Rational Mech. Anal. 4 (1960), 273-334.

12. W. Noll, Materially uniform simple bodies with inhomogeneities, Arch. Rational Mech. Anal. 27 (1967), 1-32.

13. G. E. Reyes, this volume.

Aarhus, August 1982

SYNTHETIC REASONING AND VARIABLE SETS

Gonzalo E. Reyes
Department of Mathematics
Université de Montréal
Montreal, Canada

§1 Synthetic reasoning

Whoever browses today the works of the geometers up to the end of the last century (and beginnings of ours) is struck by the type of reasoning used in their writings. It is fair to say that this "synthetic" reasoning (examples of which abound in the works of G. Darboux, S. Lie, E. Cartan and others, as well as in writings of physicists and engineers) does not fit, without violence, into the "analytic" or set-theoretical type of reasoning which has evolved since the end of the last century and has become dominant in mathematics ever since.

Let us give some examples of synthetic reasoning (which some of us, who enjoyed the benefits of a liberal engineering education, remember with nostalgia).

I (Derivatives)

Since any curve is a straight line in the infinitely small, there is a unique couple (a,b) such that

$$f(x_0+h) = a + h.b \quad \text{for all} \quad h \in D \text{ , where}$$

D = first-order infinitesimals.

Defining $f'(x_0) = b$, we conclude

$$f(x_0+h) = f(x_0) + h\, f'(x_0) \quad \text{for all} \quad h \in D$$

This formula yields the usual laws of derivation. E.g.,

$$\begin{aligned} f(g(x_0+h)) &= f(g(x_0) + h\, g'(x_0)) \\ &= f(g(x_0)) + h\, g'(x_0) f'(g(x_0)) \end{aligned}$$

since $hg'(x_0) \in D$. In other words, $(f \circ g)' = g'.f' \circ g$, the chain rule.

II (Exterior differential and Stokes' theorem)

A 1-form (resp. a 2-form) on a manifold M is an association of a number to an infinitesimal segment (resp. parallelogram) which is 0 if the segment (resp. the parallelogram) is degenerate.

If w is a 2-form and (c,h_1,h_2) is an infinitesimal parallelogram (with $h_1,h_2 \in D$)

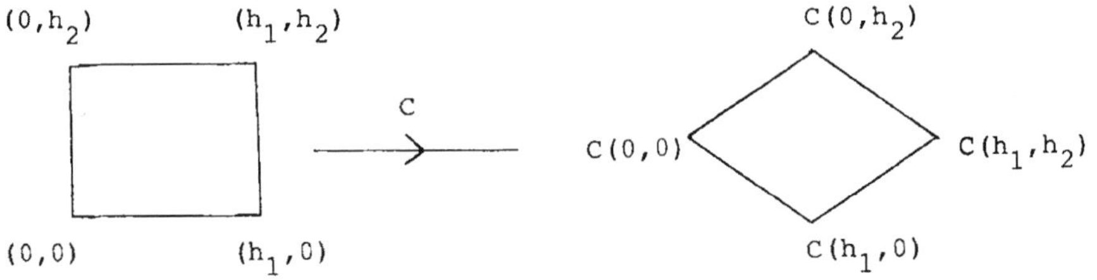

we write $\int_{(c,h_1,h_2)} w$ for the number in question

<u>Proposition</u> If w is a 1-form on manifold M there is a unique 2-form dw on M such that

$$\int_{(c,h_1,h_2)} dw = \int_{\partial(c,h_1,h_2)} w \qquad \text{for all } h_1,h_2 \in D$$

where $\int_{\partial(c,h_1,h_2)} w$ = circulation of w along (c,h_1,h_2)

<u>Proof</u> By definition of circulation,

$$\int_{\partial(c,h_1,h_2)} w = \int_{(c_I,h_1)} w + \int_{(c_{II},h_2)} w - \int_{(c_{III},h_1)} w - \int_{(c_{IV},h_2)} w$$

where $c_I(h) = c(h,0)$, $c_{II}(h) = c(h_1,h)$, $c_{III}(h) = c(h,h_2)$, $c_{IV}(h) = c(0,h)$.

By the previous argument in I, the circulation of w can be written uniquely as $a + bh_1 + ch_2 + dh_1h_2$. Furthermore, it is 0 for $h_1 = 0$ and also for $h_2 = 0$ (degeneracy condition) and this implies that

$$\int_{\partial(c,h_1,h_2)} w = h_1h_2d$$

Define $dw(c,h_1,h_2) = h_1h_2d$ to obtain the (infinitesimal) version of Stokes' theorem of the Proposition. The finite version of this theorem can be obtained by "decomposing" a parallelogram $[0,1]^2 \xrightarrow{c} M$ into infinitesimal ones and applying the infinitesimal version to each one of them.

III (Delta function)

Let $\varepsilon > 0$ be an infinitesimal and let δ be an even function such that

$$\int_{-h}^{h} \delta(t)dt = 1 \quad \text{for all} \quad h \geq \varepsilon .$$

Then, for any "standard" function f,

$$\int_{-h}^{h} f(t)\delta(t)dt - f(0) \quad \text{is infinitesimal.}$$

We notice some features of these arguments:
1) They are intrinsic; no mention is made of atlas, coordinates, etc., even when manifolds are mentioned
2) Infinitesimals are freely used and they substitute limits. Notice, however, that in I, II the infinitesimals involved are identified with certain nilpotent elements.
Obviously, these will not do for III, where a different notion of infinitesimal (closer to the one of Nonstandard Analysis) is presupposed.
3) No "logic" is used in as much as the statements concerned are equations.

The following table attempts to sum up this discussion.

Synthetic	vs Analytic
Direct manipulation of geometric objects	Manipulation of analytic representation of geometric objects
Logic at bay Use of naïve logic	Extensive use of classical logic
Extensive use of infinitesimals	limits

It was already mentioned that synthetic reasoning does not fit naturally into the set-theoretical type of reasoning. Indeed, the principle underlying the first example, namely "Any curve restricted to D is a straight line" is incompatible with classical logic: in fact, since $f'(0)$ should be completely determined by the equation

$$f(h) = f(0) + h\, f'(0) \quad \forall h \in D \ ,$$

it follows that D must contain some $h_0 \neq 0$.
Obviously, the function

$$f(h) = \begin{cases} 1 & \text{if} \quad h = h_0 \\ 0 & \text{if} \quad h \neq h_0 \end{cases}$$

cannot be developed as above.
(Otherwise, $f(h_0) = 1 = h_0 f'(0)$ and squaring both members we conclude that $1 = 0$)

This argument (due to S. Schanuel) may be considered as a modern version of Berkeley's against fluxions. There is a parallel here with Locke's critique of "abstract ideas": "... does it not require some pains and skill to form the general idea of a triangle...; for it must be neither oblique nor rectangle, neither equilateral, equicrural, nor scalenon, but all and none of these at once?"
In fact, both arguments presuppose the inconditional validity of the excluded middle and lose their destructive power when that principle is given up.

§2 Variable Sets, loci and smooth functors

In three lectures at the University of Chicago in 1967, published in A. Kock (Ed. 1979), F. W. Lawvere proposed to use the theory of variable sets (= topos theory), developed by the Grothendieck school of Algebraic Geometry, as a foundation for synthetic reasoning. This program was part of a vast research program whose aim was to provide a direct, intrinsic axiomatization of Continuum Mechanics as developed by Walter Noll and others.

Most of the "synthetic" development that has taken place along these lines, including an axiomatic foundation to justify synthetic arguments in Differential Geometry, appears in A. Kock (1981), where historical references can be found. (See also Bélair-Reyes (1982)).

In this paper, we concentrate on topos theoretic models for the axioms of Synthetic Differential Geometry (SDG, for short), especially in developments that took place after the publication of Kock's book. Missing proofs and further details will appear in a monograph in preparation.

From the point of view of variable sets, to give a continuum R

is to give a category E of "abstract" spaces and "abstract" maps, containing R as an object, together with a subcategory Z of "concrete" spaces (and "concrete" maps) subject to some relations best described as follows: we view an object F of E as a "variable set" whose elements are maps $C \to F \in E$ (where C ranges over objects of Z). Given a "concrete" map $C' \to C \in Z$ and an element of F at stage C (i.e., a map with domain C) we obtain, by composition in E, a new element of F at stage C'.
In other words, F is identified with a (contravariant) set-valued functor on Z. In a similar vein, we view a morphism $F \to G \in E$ as a map of "variable sets", which sends elements of F at stage C into elements of G at the same stage (via composition in E). This association being natural, η is thus identified with a natural transformation between the functors F and G.

These identifications amount to giving a functor

$$E \to [Z^{op}, \text{Sets}]$$

from E into the category of (contravariant) set-valued functors and natural transformations on Z, which we shall assume to be full and faithful, thus identifying the "abstract" maps of E with the corresponding natural transformations.

Notice that the naturality condition imposes restrictions on the possible "abstract" maps. In particular, Yoneda lemma tells us that "abstract" maps of E between "concrete" spaces are exactly the "concrete" maps (in Z).

Different choices of E and Z give rise to different continua: continuous reals, measurable reals, etc.

In this paper we choose Z to be the <u>category of (smooth) loci and smooth maps</u> (to be described below) and $E = [Z^{op}, \text{Sets}]$, to obtain a continuum of <u>smooth reals</u>. This continuum R will be a "variable set" whose elements, the quantities, include not only "constant quantities" such as $0, 1, \sqrt{2}, -\pi$, but "variable quantities" as well (examples of which will be the infinitesimals). Furthermore, R will be a "concrete" space (i.e., an object of Z) and this will assure that all maps from R into R are smooth. (Naturality may be seen, to some extent, as an objective counterpart to Brouwer's subjective justification of continuity principles in Intuitionism).

A (smooth) <u>locus</u> is a couple (n, I), where $n \in \mathbb{N}$ and I is an ideal of $C^\infty(\mathbb{R}^n)$ = the $\mathbb{R}$-algebra of smooth real valued functions on $\mathbb{R}^n$ (i.e., functions which are indefinitely continuously differentiable).

Intuitively, we think of (n,I) as the locus of the set of equations $f(x) = 0$, $f \in I$.

We shall write $(n,I) = [\![x \in R^n | \bigwedge_{f\in I} f(x) = 0]\!]$.

A *morphism* between (smooth) loci

$$[\![x \in R^n | \bigwedge_{f\in I} f(x) = 0]\!] \overset{\varphi}{\to} [\![x \in R^m | \bigwedge_{g\in J} g(x) = 0]\!]$$

is an equivalence class of a smooth function $\varphi: \mathbb{R}^n \to \mathbb{R}^m$ such that $g \cdot \varphi \in I$ for all $g \in J$, under the equivalence relation $\varphi \sim \psi$ iff $\pi_i \circ \varphi - \pi_i \circ \psi \in I$ for $i = 1,\ldots,m$."

Remarks

i) According to this definition, the locus $D = (1,(x^2)) = [\![x \in R | x^2 = 0]\!]$ contains lots of elements, besides the "constant quantity" $\mathbb{1} = [\![x \in R | x = 0]\!] \overset{0}{\to} D$ given by the 0 function. As an example, the "variable quantity" $D \overset{Id}{\to} D$, given by the identity function, is an element of D, "the generic first-order infinitesimal".

ii) The condition in the definition of a morphism of loci implies, but is actually stronger than, the simple minded condition

$$x \in \mathrm{Zero}(I) \Rightarrow \varphi(x) \in \mathrm{Zero}(J)$$

which would be most natural if the locus (n,I) were a set of constant quantities, rather than a variable set as intended.

As already mentioned, we let $E = [Z^{op},\mathrm{Sets}]$. The Yoneda embedding identifies Z with a (full) subcategory of E.

As examples of smooth functors, we have

i) loci themselves

The following will play important roles in what follows

$$R = (1,(0)) = [\![x \in R | \underline{0}(x) = 0]\!]$$

where $\underline{0}: \mathbb{R} \to \mathbb{R}$ is the function $\underline{0}(x) = 0 \ \forall x \in \mathbb{R}$

$$D = (1,(x^2)) = [\![x \in R | x^2 = 0]\!]$$

$$\Delta = 1, m^g_{\{0\}} = [\![x \in R | \bigwedge_{f\in m^g_{\{0\}}} f(x) = 0]\!]$$

where $f \in m^g_{\{0\}} \leftrightarrow \exists$ open $U \subseteq R$ such that $0 \in U$ and $f|_U \equiv 0$.

These loci will play the role of "reals", "first order infinitesimals" and "infinitesimals" respectively. Further loci will be introduced later on.

2) C^{∞}-manifolds (which are σ-compact)

To represent a manifold M by a locus, we embed M, by Whitney's theorem, in some $\mathbb{R}^n$ and let I be the ideal of all functions vanishing on M. The resulting locus (n,I) is, up to isomorphism in Z, independent of the embedding. For example, the 2-sphere is represented by the locus $[\![(x,y,z) \in \mathbb{R}^3 | x^2 + y^2 + z^2 - 1 = 0]\!]$.

As another example, an open $U \subseteq \mathbb{R}^n$ is represented by the locus

$$[\![(x,y) \in R^n \times R | y\chi_U(x) - 1 = 0]\!]$$

where χ_U is a "characteristic function of U" in the sense that $x \in U$ if $\chi_U(x) \neq 0$.

("Every open is Zariski open").

As suggested by the terminology of "variable sets", the category $E = [Z^{op}, \text{Sets}]$ has a set-like structure (a topos). The following Proposition isolates a few of these properties which are enough for the purpose of this paper.

Proposition $E = [Z^{op}, \text{Sets}]$ is a cartesian closed category.

Proof. Essentially obvious, once the meaning is understood: E has finite products, equalizers, terminal object and exponentials. As to products: if $F, G \in E$, we define $F \times G$ by stipulating that an element at stage C is a couple consisting of an element of F and an element of G at that stage. We write

$$\frac{C \to F \times G}{C \to F, \; C \to G}$$

As to terminal object: $\mathbb{1} = [\![x \in \mathbb{R} | x = 0]\!]$

As to exponentials:

$$\frac{C \to F^G}{C \times G \to F}$$

To finish the proof, just check the universal properties of products, exponentials, etc.

E.g. $\dfrac{H \to F^G}{H \times G \to F}$ for all $F,G,H \in E$.

Remark Notice that $(n,I) \times (m,J) = (n+m,(I,J))$.

The relations between the categories introduced so far can be described by means of several functors which we proceed to define.

Let $s: \mathcal{M} \to Z$ be the functor given by Whitney's embedding theorem (mentioned above), from the category $\mathcal{M}$ of σ-compact C^∞-manifolds (possibly with corners) into Z.

Proposition s is full and faithful and preserves transversal pull-backs of $\mathcal{M}$.

Remark For manifolds without boundary, this notion is essentially the same as transversal intersection; for others see Quê-Reyes (1982).

Let $\Gamma: E \to \text{Sets}$ be the functor of "global sections" defined by $\Gamma(F) = E(\mathbf{1},F)$ = natural transformations from $\mathbf{1}$ to F (with the obvious action on morphisms).

Furthermore, we define

$\Delta: \text{Sets} \to E$, the constant functor, by

$\Delta(X)(C) = X$ for all $C \in Z$ and

$\Delta(X)(C' \overset{\varphi}{\to} C) = id_X$ for all $\varphi \in Z$.

The functor Γ is a left adjoint to $\Gamma(\Delta \dashv \Gamma)$ in the sense that there is a bijection between morphisms.

$$\frac{\Delta(X) \to F \in E}{X \to \Gamma(F) \in \text{Sets}}$$

We have a further functor

$B: \text{Sets} \to E$, defined by

$B(X)(C) = \text{Sets}(\Gamma(C),X) = X^{\Gamma(C)}$

(with obvious action on functions)

The fundamental relations between these functors are

$\Delta \dashv \Gamma \dashv B$

We have the following diagram

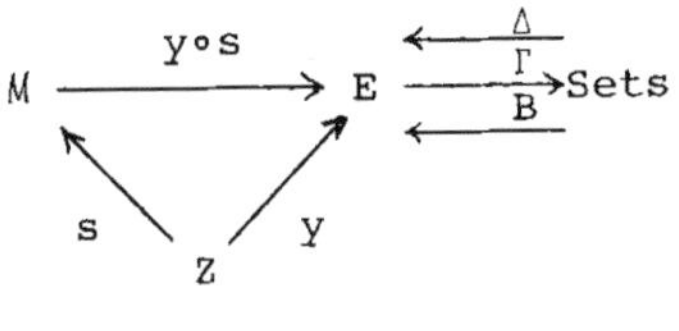

Remarks i) In terms of the logic of E, the existence of the functor B implies the validity of the intuitionistic principles

$$E \Vdash \exists x \varphi \Rightarrow E \Vdash \varphi(a) \quad \text{for some } a$$

$$E \Vdash \sigma \vee \tau \Rightarrow E \Vdash \sigma \quad \text{or} \quad E \Vdash \tau$$

ii) The endofunctor $(-)^D$ of E which obviously has $(-) \times D$ as a left adjoint has also a right adjoint $(-)_D$:

$$(-) \times D \dashv (-)^D \dashv (-)_D$$

This expresses the fact that the infinitesimal space D is an "atom" (internally projective).

§3 A model

We give here an interpretation or model, obtained via the Yoneda embedding, of the fundamental notions of synthetic reasoning in terms of classical notions of smooth functions and ideals of such functions. This interpretation is obtained by giving explicit descriptions for the elements of R, D, R^D, etc. in E at the stage (n,I).

E.g.

$$\frac{(n,I) \xrightarrow{f} R^D}{\dfrac{(n,I) \times D \xrightarrow{f} R}{\dfrac{(n,I) \times (1,t^2)) \xrightarrow{f} R}{(n+1,(I,t^2)) \xrightarrow{f} R}}}$$

Definition of exponential

Definition of D

Computation of products

In other words, we have obtained:

A function in R^D at stage (n,I) is an equivalence class $f(x,t)$ mod $(I,(t^2))$.

The following clauses are obtained in the same way:

(1) A real at stage (n,I) is an equivalence class $f(x)$ mod I, where $f \in C^\infty(\mathbb{R}^n)$. (Notice that a representative of a real is an element of $\mathbb{R}$ which depends smoothly on the parameter x)

(2) $0,1,+,\cdot$ at stage (n,I):

$$0 = 0 \bmod I$$
$$1 = 1 \bmod I$$

$$(f \bmod I) + (g \bmod I) = (f+g) \bmod I$$
$$(f \bmod I) \cdot (g \bmod I) = (f.g) \bmod I$$

(3) A first order infinitesimal (i.e. a real in D) at the stage (n,I) is a class f mod I, with $f^2 \in I$.

(4) A function in R^R at stage (n,I) is a class $F(x,t)$ mod I^* where $F \in C^\infty(\mathbb{R}^n \times \mathbb{R})$, I^* is the ideal in $C^\infty(\mathbb{R}^n \times \mathbb{R})$ generated by $\{f \cdot \pi \mid f \in I\}$ and $\pi: \mathbb{R}^n \times \mathbb{R} \to \mathbb{R}^n$ is the canonical projection. (Notice

that a representative of a function is then a smooth function of t which depends smoothly on the parameter x).

(5) A function in R^D at stage (n,I) is a class $F(x,t) \bmod (I,(t^2))$.

(6) The evaluation of a function at a real at stage (n,I) is a class $F(x,f(x)) \bmod I$, where $F(x,t) \bmod I^*$ is the function and $f \bmod I$ is the real in question.

Already at this level we can justify the reasoning underlying Example I

Proposition (Kock-Lawvere Axiom holds). Any function in R^D is (uniquely) a straight line.

Proof Let $F(x,t) \in C^\infty(\mathbb{R}^n \times \mathbb{R})$. Using the Fundamental Theorem of Calculus twice (in the version of Hadamard), we conclude the existence of $G(x,t) \in C^\infty(\mathbb{R}^n \times \mathbb{R})$ such that

$$F(x,t) = F(x,0) + t\frac{\partial F}{\partial t}(x,0) + t^2 G(x,t)$$

$$\text{Hence, } F(x,t) = F(x,0) + t\frac{\partial F}{\partial t}(x,0) \bmod (I,(t^2))$$

$$\text{Let } a = F(x,0) \bmod I$$

$$b = \frac{\partial F}{\partial t}(x,0) \bmod I$$

$$\text{Therefore, } f = a + hb \quad \forall h \in D .$$

We can extend our dictionary to accomodate strict order and pre-order relations in a natural way: let $R_{>0} = s(\mathbb{R}_{>0})$ and $R_{\geq 0} = s(R_{\geq 0})$. We define a real at stage (n,I) to be positive (resp. non-negative) if the corresponding morphism $(n,I) \to R$ factors through $R_{>0} \to R$ (resp. $R_{\geq 0} \to R$).

We can easily prove:

(7) A positive real at stage (n,I) is a class $f \bmod I$ such that, for some function $g \in C^\infty(\mathbb{R}^n)$ $\chi(f(x)).g(x) = 1 \bmod I$, where $\chi(x) \neq 0$ iff $x > 0$ (I.e., χ is a "characteristic function" of $\mathbb{R}_{>0}$)

(8) A non-negative real at stage (n,I) is a class $f \bmod I$ such that

$$\varphi(f(x)) \in I, \text{ for all } \varphi \in m^\infty_{\mathbb{R}_{\geq 0}}$$

(We use the following notation for $X \subseteq \mathbb{R}^n$: m^∞_X is the ideal of flat functions on X, i.e., the ideal of functions ff such that $D^\alpha f|_X \equiv 0$, for any multi-index α).

Proposition The strict order and pre-order relations defined above are compatible with the algebraic structure of R and among themselves (i.e., a positive real is non-negative). Furthermore nilpotent elements are non-negative.

Proof For the strict order relation, see Joyal-Reyes (1982). For the other, the proof uses the following.

Proposition If $X \subseteq \mathbb{R}^n$, $Y \subseteq \mathbb{R}^m$, are closed, then $m^\infty_{X\times Y} = (m^\infty_X, m^\infty_Y)$. For the proof (of this last Proposition) see Quê-Reyes (1982).

Let us check that the sum of two non-negative reals at stage (n,I) is again non-negative. Let $f \bmod I$, $g \bmod I$ be the reals in question and let $\varphi \in M^\infty_{\mathbb{R}\geq 0}$. Clearly $\varphi(u+v) \in m^\infty_{\mathbb{R}_{\geq 0} \times \mathbb{R}_{\geq 0}}$ and by the previous proposition,

$$\varphi(u+v) = F(u,v)\theta(u) + G(u,v)\Psi(v), \quad \text{where } \theta, \Psi \in m^\infty_{\mathbb{R}_{\geq 0}} .$$

Substituting $f(x)$ for u and $g(x)$ for v, we conclude that $\varphi(f(x)+g(x)) \in I$.

The justification of the finite version of Stokes' theorem in Example II depends on the existence of primitives of functions in $R^{[0,1]}$.

Let us define $[0,1]$ as $s\{x \in \mathbb{R} \mid 0 \leq x \leq 1\}$.

Proposition (Integration holds in our interpretation)

$$\forall f \in R^{[0,1]} \; \exists! g \in R^{[0,1]} \; g' \equiv f \wedge g(0) = 0 .$$

Proof Using Yoneda as before, it turns out that a function in $R^{[0,1]}$ at stage (n,I) is a class $f = F(x,t) \bmod (I, m^\infty_{[0,1]})$, where $F \in C^\infty(\mathbb{R}^n \times \mathbb{R})$. Define $g = \int_0^t F(x,u)\,du \bmod (I, m^\infty_{[0,1]})$. To check that g is well defined, use once more the Proposition on ideals of flat functions.

We will only touch upon the justification of the reasoning underlying Example III; further details can be found in Reyes 1982 as well as in the forthcoming monograph mentioned in §2.

We already mentioned that

$$\Delta = (1, m^g_{\{0\}}) = [\![x \in \mathbb{R} \mid \bigwedge_{f \in m^g_{\{0\}}} f(x) = 0]\!] \quad \text{plays the role of}$$

infinitesimals. To justify this assertion we notice that $\Delta \to R$ factors through $s(U)$ for every open $U \subset \mathbb{R}$ which contains 0. I.e., the elements of Δ are smaller than all constant positive reals and larger than all negative reals. The smooth functor

$$\mathbb{I} = \Delta \cap s(\mathbb{R} - \{0\})$$

plays the role of the "invertible infinitesimals".

We consider, as the new model, the "slash" category $E/\mathbb{I}$ whose objects are morphisms $F \to \mathbb{I}$ (with $F \in E$) and whose morphisms are commutative triangles

$$\begin{array}{ccc} F & \to & G \\ & \searrow \quad \swarrow & \\ & \mathbb{I} & \end{array}$$

Notice that we have a functor

$$E \to E/\mathbb{I}$$

which sends X into $X \times \mathbb{I} \overset{\pi_2}{\to} \mathbb{I}$. This allows to construct the composite functor

$$M \overset{s}{\to} E \to E/\mathbb{I}$$

which is not full (since there are no invertible infinitesimals in $\mathbb{R}$!) , but still faithful.
Furthermore, it preserves transversal pull-backs.

To construct our function δ in $E/\mathbb{I}$, let $\delta_0 \in C^\infty(\mathbb{R})$ be any even function as shown below

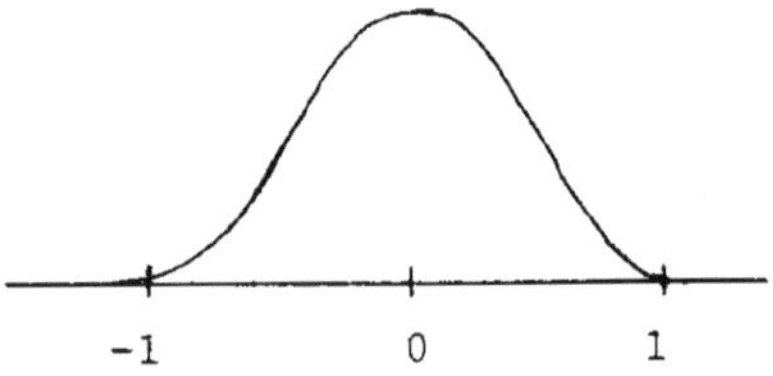

with area $\int_{-1}^{1} \delta_0(t)dt = 1$. We define

$$\delta(t) = \frac{1}{\varepsilon} s(\delta_0)(\varepsilon t)$$

where ε is the "generic invertible infinitesimal" $\varepsilon : \mathbb{I} \to \mathbb{I} \times \mathbb{I}$ (the diagonal) in $\varepsilon/\mathbb{I}$.

Remark To simplify our exposition, only categories E of the form $[Z^{op},Sets]$ were considered, although they are not quite adequate as models of synthetic reasoning. Indeed, the functor $M \to E$ does not preserve open coverings; equivalently, R does not have good algebraic properties (in particular it is not even a local ring).

Using standard techniques of topos theory (the so-called Grothendieck topologies) it is possible to construct categories of sheaves $Sh(Z) \to E$ and a factorization

$$M \to Sh(Z) \to E$$

Such that the first functor preserves the required unions (e.g. finite or countable ones). (See Dubuc (1979), (1980) as well as the monograph mentioned in §2).
The object R of smooth reals in Sh(Z) becomes a local ring, which is even real closed in an appropriate sense (see Joyal-Reyes (1982)).

§4 Acknowledgments

The research reported in this paper was partially supported by the Natural Sciences and Engineering Research Council of Canada, the Ministère de l'Education du Gouvernement du Québec and the Australian Research Grants Committee. This support is gratefully acknowledged.

Most of the results mentioned in this paper were presented in series of lectures at the Pontificia Universidad Católica de Chile (Santiago, Dec. 81-Jan. 82), the Université Catholique de Louvain (Louvain-la-Neuve, April-May 82) and the University of Sydney (Sydney, June-July 82). We would like to thank Rolando Chuaqui, Francis Borceux and Max Kelly for making these visits possible. It was a unique opportunity for fruitful and pleasant contacts with people in three continents.

The actual presentation of this paper has been much influenced by Lawvere's talks at this conference.

REFERENCES

[1] M. Artin, A. Grothendieck, and J. L. Verdier, (1972) Théorie des topos et cohomologie étale des schémas (SGA 4), Vol. 1, Lecture Notes in Math. 269, Springer-Verlag.

[2] L. Bélair and G. E. Reyes (1982), Calcul Infinitésimal en Géometrie Différentielle Synthétique, to appear in Proc. First Symposium of Chilean Mathematicians (Valparaiso 1981).

[3] E. Cartan (1951), Leçons sur la Géometrie des espaces de Riemann, Gauthier-Villars.

[4] G. Darboux (1887-1896), Leçons sur la théorie des surfaces, Gauthier-Villars.

[5] E. J. Dubuc (1979), Sur les modèles de la Géometrie Différentielle Synthétique, Cahiers de Top. et Géom. Diff. 20 (1979), 231-279.

[6] E. J. Dubuc, C^{∞}-schemes, American Journ. Math. 103, (1981).

[7] A. Joyal and G. E. Reyes, Separable real closed local rings, to appear in Proc. of the V Symposium of Logic in Latin America (Botatá 1981).

[8] A. Kock (Ed. 1979), Topos theoretic methods in geometry, Aarhus Math. Inst. Var. Publ. Series No. 30.

[9] A. Kock (1981), Synthetic Differential Geometry, London Mathematical Society Lecture Notes Series 51, Cambridge University Press.

[10] S. Lie (1876), Allgemeine Theorie der partiellen Differentialgleichungen erster Ordnung, Math. Ann. 9 (186), 245-296.

[11] W. Noll (1974), The Foundations of Mechanics and Thermodynamics, Springer-Verlag.

[12] Ngo van Quê and G. E. Reyes (1982), Smooth functors and Synthetic Calculus, to appear in the Proc. Brouwer Centenary Conference.

[13] G. E. Reyes (1982), Types of infinitesimals, to appear.

[14] A. Robinson (1966), Non Standard Analysis, North-Holland.

RECENT RESEARCH ON THE FOUNDATIONS OF THERMODYNAMICS*

Bernard D. Coleman and David R. Owen
Department of Mathematics
Carnegie-Mellon University
Pittsburgh, PA 15213

The principles of thermodynamics have found application in many branches of science. These principles have been employed to understand the efficiency of heat engines, the electromotive force of galvanic cells and thermal junctions, the dependence of chemical equilibrium on temperature and pressure, the properties of phase transitions, and the thermomechanics of continuous bodies.

Although thermodynamics is the science of heat and temperature, its principles are often usefully applied to experiments in which heat is not flowing (e.g., those involving poor thermal conductors or insulated reaction chambers) or others in which temperature is not changing (because, say, the object under study is a good thermal conductor in contact with an isothermal environment). One recognizes a thermodynamical argument by its reference to consequences of either the first or the second law. Every student of physics or chemistry has been taught that the first law is an assertion about the balance of heat and work, and that the second law is an assertion about the rate of increase of entropy that, in some sense, is equivalent to a denial of the existence of certain perpetual motion machines, or to a denial of the existence of cycles in which heat is absorbed at some temperatures without emission at others, or to an assumption about the sign of the sum over a cycle of the ratio of the heat absorbed to the temperature at which it is absorbed.[1)]

We here describe some recent work toward a precise formulation of the second law as a general principle whose implications can be derived with rigor.[2)] We do not believe that the results of this work can be

---

[1)] In their explanation of the laws of thermodynamics, particularly the second law, the texts on the subject tend to be obscure, not because the principles sought fail to have the generality claimed, but because of the absence of a mathematical language that permits expression of the principles at that level of generality.

[2)] This brief survey is confined to the study of certain forms of the second law. No attempt is made to present a complete set of axioms for all of thermodynamics. The first law and the concept of work are not discussed. For accounts of the early development of the science of thermodynamics and the discovery of the first and second laws, see the books of Truesdell [1980,2] and Truesdell & Bharatha [1977,1].

[*)] A later version of this paper appeared as an appendix in C. Truesdell's Rational Thermodynamics, second edition, Springer Verlag [1984].

dismissed as "mere axiomatics". The development in the 1960's of the thermodynamics of materials with memory raised questions whose resolution required a careful examination of the mathematical foundations of thermodynamics. In its original presentation, the theory of the thermodynamics of materials with fading memory rested on the Clausius-Duhem inequality, i.e., on the assumption that for each substance there is a function of state,[3] called the entropy, whose difference at two states dominates the ratio of the heat absorbed to the absolute temperature along each process taking one state to the other. The question was raised: does each substance have such a function of state with the properties of regularity needed to derive the now known consequences of the Clausius-Durhem inequality? Of course, the question is meaningful only if one has a statement of the second law that does not presuppose the presence of entropy as a function of state. A statement of this type can be obtained by making mathematical the ideas behind the familiar assertion that *the sum along a cycle of the ratio of the heat gained to the absolute temperature at which it is gained cannot be positive*. However, to be useful for materials with gradually fading memory and for other substances with few nontrivial cycles, the statement must be formulated in such a way that it has meaning for "approximate cycles". We have obtained such a formulation of the second law and have used it to study various questions, including the existence, uniqueness, and regularity of entropy functions.

In this recent work [1974,1][1975,1], a careful distinction is made between the general structure of thermodynamical systems and the equations defining special classes of systems.[4] The concept of a *system* employed is one in which a system is a pair (Σ,Π) of sets with the following mathematical structure: Σ is a topological space whose elements are the *states*; Π is the set of *processes*; associated with each process is a continuous function ρ_P, mapping a non-empty open subset $\mathcal{D}(P)$ of Σ onto a subset $\mathcal{R}(P)$ of Σ ; ρ_P is called the *transformation induced by* P and its value at a state σ in $\mathcal{D}(P)$ is denoted by $\rho_P\sigma$; to each pair (P'',P') of processes for which $\mathcal{R}(P')$ intersects $\mathcal{D}(P'')$ there is assigned a process $P''P'$ called the *process resulting from the successive application of* (first) P'

[3] For a material with fading memory, a *state* can be identified with an appropriate *history*.

[4] It is common for the older literature to obscure such a distinction or to employ two languages: the language of mathematics for the treatment of special systems and examples, and a non-mathematical language, reminiscent of metaphysics, for discussion of the general principles of thermodynamics.

and (then) P". It is assumed that $(\underset{\sim}{I})$ for each σ in Σ, the set of states accessible from σ, i.e., the set of states of the form $\rho_P\sigma$ with P in Π, is dense in Σ, and $(\underset{\sim}{II})$ if $P''P'$ is the result of the successive application of P' and P'', then the transformation $\rho_{P''P'}$ induced by $P''P'$ is the composition of $\rho_{P'}$ and $\rho_{P''}$, i.e., is the function defined on $\mathcal{D}(P''P') = \rho_{P'}^{-1}(\mathcal{D}(P''))$ by the equation $\rho_{P''P'}\sigma = \rho_{P''}\rho_{P'}\sigma$.

The mathematical concept that renders precise and general the idea of a "sum along a process" is that of an action. An action is a function that assigns a number $\alpha(P,\sigma)$ to each pair (P,σ) with P in Π and σ in $\mathcal{D}(P)$; $\alpha(P,\sigma)$ is called the supply of α on going from σ to $\rho_P\sigma$ via the process P. Two properties are required of an action: (i) additivity in the sense that if P is the result of the successive application of P' and P'', then for each σ in $\mathcal{D}(P)$ the supply of α obtained by going from σ to $\rho_P\sigma$ via P is the sum of the supplies of α obtained by going from σ to $\rho_{P'}\sigma$ via P' and from $\rho_{P'}\sigma$ to $\rho_P\sigma = \rho_{P''}\rho_{P'}\sigma$ via P'', i.e.,

$$\alpha(P,\sigma) = \alpha(P',\sigma) + \alpha(P'',\rho_{P'}\sigma) , \tag{1}$$

and (ii) continuity in the sense that for each P in Π, the function $\alpha(P,\cdot)$ is continuous on $\mathcal{D}(P)$.

A process P and a state σ° are said to form a cycle (P,σ°) if σ° is in $\mathcal{D}(P)$ and $\rho_P\sigma^\circ = \sigma^\circ$. One may consider taking the second law to be the assertion that an appropriate action $\mathcal{S}$ (which, of course, must be specified) is not positive when its argument is a cycle, i.e., is such that

$$\rho_P\sigma^\circ = \sigma^\circ \Longrightarrow \mathcal{S}(P,\sigma^\circ) \leq 0 . \tag{2}$$

For materials with gradually fading memory, the class of cycles (P,σ°) is too small for (2) to have the full implications expected of the second law. To obtain an extension of (2) to "approximate cycles", we have employed the following concept: We say that $\mathcal{S}$ has the Clausius property at a state σ° if, for each $\varepsilon > 0$, σ° has a neighborhood $\mathcal{O}_\varepsilon(\sigma^\circ)$ for which

$$\rho_P\sigma^\circ \in \mathcal{O}_\varepsilon(\sigma^\circ) \Rightarrow \mathcal{S}(P,\sigma^\circ) < \varepsilon . \tag{3}$$

It is clear that if (3) holds and $\rho_P\sigma^\circ = \sigma^\circ$, we have $\mathcal{S}(P,\sigma^\circ) < \varepsilon$ for every $\varepsilon > 0$, and hence (2) holds; i.e., if $\mathcal{S}$ has the Clausius property at σ°, then $\mathcal{A}(P,\sigma^\circ)$ is not positive when (P,σ°) is a cycle.

If an action has the Clausius property at a state σ°, then it has the property at each state in a set Σ° that is dense in Σ and contains all states σ that are accessible from σ°;[5)] this set Σ° may be defined as follows:

For each state σ° let $\mathfrak{S}(\sigma^\circ,\sigma)$ be the collection of all the open subsets $\mathfrak{O}$ of Σ that contain σ and are such that the sets $\mathcal{A}\{\sigma^\circ \to \mathfrak{O}\}$, defined by

$$\mathcal{A}\{\sigma^\circ \to \mathfrak{O}\} = \{\mathcal{A}(P,\sigma^\circ) \mid P \in \Pi,\ \rho_P\sigma^\circ \in \mathfrak{O}\}, \tag{4}$$

are individually bounded above, i.e., have

$$\sup \mathcal{A}\{\sigma^\circ \to \mathfrak{O}\} < \infty ; \tag{5}$$

Σ° is the set of states for which

$$m(\sigma^\circ,\sigma) := \inf_{\mathfrak{O}\in\mathfrak{S}(\sigma^\circ,\sigma)} \sup \mathcal{A}\{\sigma^\circ \to \mathfrak{O}\} \tag{6}$$

is finite, i.e.,

$$\Sigma^\circ = \{\sigma \mid \sigma \in \Sigma,\ m(\sigma^\circ,\sigma) > -\infty\}. \tag{7}$$

In [1974,1] we interpreted the second law as the statement that <u>a particular action $\mathcal{A}$ has the Clausius property at least at one state</u>. To prove that such a statement implies the existence of an entropy function that enters a relation with the form of the Clausius-Duhem inequality, we there introduced the concept of an upper potential.

A real valued function S on a dense subset $\mathfrak{S}$ of Σ is called an <u>upper potential</u> for an action $\mathcal{A}$ if for each pair of states σ_1,σ_2 in $\mathfrak{S}$ and each $\varepsilon > 0$ there is a neighborhood $\mathfrak{O}_\varepsilon(\sigma_1,\sigma_2)$ of σ_2 such that whenever $\rho_P\sigma_1$ is in $\mathfrak{O}_\varepsilon(\sigma_1,\sigma_2)$ there holds

$$S(\sigma_2) - S(\sigma_1) > \mathcal{A}(P,\sigma_1) - \varepsilon . \tag{8}$$

In the special case in which σ_2 is accessible from σ_1, i.e., in which σ_2 has the form $\sigma_2 = \rho_P\sigma_1$, this relation holds for all $\varepsilon > 0$

5) [1974,1] Thm. 3.1.

and hence implies that the supply of $\mathscr{S}$ on going from σ_1 to σ_2 is dominated by the difference $S(\sigma_2) - S(\sigma_1)$:

$$S(\rho_P \sigma_1) - S(\sigma_1) \geq \mathscr{S}(P, \sigma_1) \ . \tag{9}$$

It is easily seen that an action that has an upper potential has the Clausius property at each state in the domain of the upper potential. Although far less trivial to show, the converse is also true: *the assumption that there are states at which* $\mathscr{S}$ *has the Clausius property implies that* $\mathscr{S}$ *has an upper potential, in fact, one that is upper semicontinuous.*[6] If we identify $\mathscr{S}(P, \sigma)$ with the sum of the heat added divided by the temperature at which it is added as the system is taken from the state σ to the state $\rho_P \sigma$ by the process P, then in (9) the upper potential S is playing the role played by entropy in the Clausius-Duhem inequality. Thus, the existence of an upper potential for $\mathscr{S}$ is tantamount to the existence of entropy as a function of state.

Our construction of an entropy function employs the observation that *if* $\mathscr{S}$ *has the Clausius property at* σ° *and if* S° *is defined on* Σ° *by*

$$S^\circ(\sigma) = m(\sigma^\circ, \sigma) \ , \tag{10}$$

then not only is the domain Σ° *of* S° *dense in* Σ, *but* S° *is an upper potential for* $\mathscr{S}$ *and is upper semicontinuous on* Σ°.

Under the assumptions stated up to this point, we can say no more about the regularity of upper potentials for $\mathscr{S}$ than that there is one that is semicontinuous. This should not be surprising, for we have so far assumed very little about systems and actions. At this level of generality, the collection Σ of states has a topology but not the vector-space or manifold structure required to make meaningful the concept of a differentiable function on Σ. When more is assumed about the system (Σ, Π) and the action $\mathscr{S}$, one expects to be able to prove more about entropy as a function of state.

We have not yet assumed any special properties for the state σ° with which we start when we construct Σ° as shown in equations (4)-(7). We would like to take this state as a "standard state" and be able to normalize entropy functions S on Σ° so that

$$S(\sigma^\circ) = 0 \ . \tag{11}$$

---

6) [1974,1] Thm. 3.1.

Although the hypotheses we have made so far imply that Σ° is dense in Σ, they are not strong enough to imply that σ° is in Σ°. It does suffice, however, to assume that this selected state σ° is *equilibrated with respect to* $\mathscr{S}$ in the sense that there is at least one process P° in Π for which

$$\rho_{P^\circ}\sigma^\circ = \sigma^\circ \quad \text{and} \quad \mathscr{S}(P^\circ,\sigma^\circ) = 0 . \tag{12}$$

In fact, we have the following theorem:[7)] *Suppose that* σ° *is equilibrated with respect to* $\mathscr{S}$ *and is a state at which* $\mathscr{S}$ *has the Clausius property. Then* **(1)** σ° *is in* Σ°; **(2)**

$$m(\sigma^\circ,\sigma^\circ) = 0 , \tag{13}$$

and hence the upper potential S° *defined in* (10) *vanishes at* σ°, i.e.,

$$S^\circ(\sigma^\circ) = 0 ; \tag{14}$$

moreover **(3)** S° *is the smallest entropy function that is normalized in this way: if* S *is an upper potential for* $\mathscr{S}$ *that is defined on* Σ° *and obeys* (11), *then for each state* σ *in* Σ°,

$$S^\circ(\sigma) \leq S(\sigma) . \tag{15}$$

If, in addition, $m(\sigma,\sigma^\circ)$ *[defined by interchanging the roles of* σ *and* σ° *in the relations* (4)-(6)*] is finite for each* σ *in* Σ°, *then the function* S_o *defined on* Σ° *by*

$$S_o(\sigma) = -m(\sigma,\sigma^\circ) \tag{16}$$

also is an upper potential for $\mathscr{S}$ *and not only obeys the normalization* (11), *but is the largest entropy function that does;* i.e., *each upper potential for* $\mathscr{S}$ *that is defined on* Σ° *and obeys* (11) *has the bounds:*

$$S^\circ(\sigma) \leq S(\sigma) \leq S_o(\sigma) . \tag{17}$$

The set of entropy functions on Σ° normalized according to (11)

7) [1975,1] §3. In [1974,1] §7 we show that a weaker hypothesis, namely that σ° be a relaxed state, suffices for the theorem.

is convex: if S_1 and S_2 are two such entropy functions, then so also is each function of the form $\alpha S_1 + (1-\alpha)S_2$, $0 < \alpha < 1$. We have just observed that if $\mathcal{A}$ has the Clausius property at σ° , if σ° is equilibrated with respect to $\mathcal{A}$ at σ° , and if $m(\sigma,\sigma^\circ)$ is finite whenever $m(\sigma^\circ,\sigma)$ is, then S° is the maximal and $S_\circ$ is the minimal element of this convex set of normalized entropy functions. Clearly, then, this set reduces to singleton if and only if $S^\circ = S_\circ$. That is, under these hypotheses about the reference state σ° , *in order that there be only one entropy function* S *on* Σ° *obeying* (11) *it is necessary and sufficient that*

$$m(\sigma^\circ,\sigma) + m(\sigma,\sigma^\circ) = 0 . \tag{18}$$

This condition, although met by elastic materials and viscous materials, is not met in general. Among the exceptions are certain elastic-plastic materials, materials with fading memory, and certain materials with internal state variables.

We have seen that if one takes the second law to be the assertion that an action $\mathcal{A}$ has the Clausius property, one can deduce the existence of entropy as a function of state and obtain information about the regularity and uniqueness of entropy. We have left open the questions: (i) Which of the many actions one can formulate for a system should be assumed to have the Clausius property? (ii) Can information about the form of $\mathcal{A}$ be deduced from a statement of the second law that makes precise an assertion to the effect that there can be no cycles in which heat is only absorbed? (iii) In what sense is "absolute temperature" a distinguished measure of hotness?[8]

Recent papers of Serrin have shed light on these questions. Basic to his theory [1979,2,3] is the concept of the *hotness manifold* $\mathcal{H}$, introduced by Mach [1896,1] and assumed by Serrin to be a continuous, oriented, one-dimensional manifold whose points L are called *levels of hotness*, or, for short, *hotnesses*. It is assumed that the

8) There is a growing literature devoted to questions related to these. We mention here a few recent contributions. Boyling [1972,1] has discussed the construction of entropy and absolute temperature from axioms of the form proposed by Caratheodory. Truesdell & Bharatha [1977,1] and Truesdell [1979,4] have clarified and extended Carnot, Reech, and Kelvin's studies of ideal (reversible) systems. Serrin's research [1979,2,3] and our own research done with him [1981,1] extend and render mathematical ideas expressed by Kelvin and Clausius. Recent papers expressing a similar point of view and showing points of contact with the research described below are those of Šilhavý [1980,1],[1982,3], Feinberg & Lavine [1982,1], and Owen [1982,2].

orientation of $\mathcal{H}$ induces a total strict order "$\prec$" on hotnesses, with "$L_1 \prec L_2$" read "L_1 is a lower level of hotness than L_2", or "L_1 is below L_2", or "L_2 is above L_1". Serrin's theory [1979,2,3] does not rest on a concept of "state", but does refer to objects that we may here identify with <u>cycles</u>, i.e., with pairs (P,σ) in $\Pi \times \Sigma$ with σ in $\mathcal{D}(P)$ and $\rho_P\sigma = \sigma$. Let us define a <u>classical thermodynamical system</u> to be a set $\mathbb{P}_c$ of cycles and a real-valued function Q on $\mathbb{P}_c \times \mathcal{H}$, called the <u>accumulation function</u>; the value $Q(\mathcal{P},L)$ at a point $(\mathcal{P},L)$ in $\mathbb{P}_c \times \mathcal{H}$ is called the <u>net heat absorbed by the system at levels of hotness at or below</u> L <u>in the cycle</u> $\mathcal{P}$. It is assumed that $Q(\mathcal{P},L)$ varies only over a bounded interval in $\mathcal{H}$ in the sense that for each cycle $\mathcal{P}$ there are levels of hotness $L^\ell = L^\ell(\mathcal{P})$, $L^u = L^u(\mathcal{P})$, with $L^\ell \prec L^u$, such that

$$\left.\begin{aligned} Q(\mathcal{P},L) &= 0 \quad \text{for} \quad L \prec L^\ell , \\ Q(\mathcal{P},L) &= Q(\mathcal{P},L^u) \quad \text{for} \quad L^u \preceq L . \end{aligned}\right\} \tag{19}$$

For each $\mathcal{P}$, the function $Q(\mathcal{P},\cdot)$ generates a finitely additive set function $q_{\mathcal{P}}$ for $\mathcal{H}$ whose value $q_{\mathcal{P}}(I) = Q(\mathcal{P},L_2) - Q(\mathcal{P},L_1)$ on the set $I = \{L \mid L_1 \prec L \preceq L_2\}$ is the <u>net heat absorbed by the system at levels of hotness in</u> I, (i.e., <u>above</u> L_1 <u>and at or below</u> L_2); (19) implies that this set function has compact support. The number $Q(\mathcal{P},L^u)$ is called the <u>overall net gain of heat</u> (by the system) <u>in the cycle</u> $\mathcal{P}$.

Serrin develops a language for discussing the effect of the operation of two or more systems or the repeated operation of a single system: if $\mathcal{S}' = (\mathbb{P}_c',Q')$ and $\mathcal{S}'' = (\mathbb{P}_c'',Q'')$ are two thermodynamical systems (which may or may not be the same), their <u>union</u> $\mathcal{S}' \oplus \mathcal{S}''$ is the thermodynamical system $\mathcal{S} = (\mathbb{P}_c',Q)$ with

$$\mathbb{P}_c = \mathbb{P}_c' \times \mathbb{P}_c'' \tag{20}$$

and

$$Q((\mathcal{P}',\mathcal{P}''),L) = Q'(\mathcal{P}',L) + Q''(\mathcal{P}'',L) \tag{21}$$

for each pair $(\mathcal{P}',\mathcal{P}'')$ in $\mathbb{P}_c$ and each L in $\mathcal{H}$.

Serrin assumes that a collection $\mathbb{U}$ of classical thermodynamical systems, closed under the union operation, has been given. His statement of the second law is: <u>If</u> $\mathcal{P}$ <u>is a cycle of a system</u> $(\mathbb{P}_c,Q)$ <u>in</u> $\mathbb{U}$ <u>with</u> $Q(\mathcal{P},L) \geq 0$ <u>for every</u> L, <u>then</u> $Q(\mathcal{P},L) = 0$ <u>for every</u> L. In other words, in a cycle for which the net heat absorbed at or below each hotness level is not negative, the accumulation function is identically zero and, in particular, the overall net gain of heat in that cycle is zero.

To show that his statement of the second law permits the construction of a function $\mathscr{s}$ obeying (2), Serrin assumes that the collection $\mathfrak{U}$ of classical thermodynamical systems contains at least one special system that is the mathematical embodiment of an elastic or viscous substance. His proofs take their simplest form if the distinguished systems are <u>ideal gases</u>; these are systems for which each sycle $\mathcal{P}$ can be represented as a closed (oriented) curve $\mathcal{L}_{\mathcal{P}}$ in the first quadrant of a coordinate plane with one coordinate, V, interpreted as the volume of the gas, and the other, θ, a coordinate indicating the level of hotness L in the gas. The number θ is related to L by a strictly increasing, positive-valued, continuous function φ on the manifold $\mathcal{H}$. (Use of the coordinate system φ on $\mathcal{H}$ corresponds to measurement of hotness with an "ideal gas thermometer".) For each process $\mathcal{P}$ of an ideal gas and each level $\bar{L}$ of hotness, $Q(\mathcal{P},\bar{L})$ equals the integral of a differential form, $c(V,\theta)d\theta + p(V,\theta)dV$, over the portion $\mathcal{L}_{\mathcal{P}}(\bar{L})$ of $\mathcal{L}_{\mathcal{P}}$ on which $L \preceq \bar{L}$, i.e., on which the coordinate $\theta = \varphi(L)$ is equal to or less than $\bar{\theta} = \varphi(\bar{L})$; p is the pressure in the gas and is given by the formula,

$$p(V,\theta) = \frac{r\theta}{V}\ . \tag{22}$$

with r a constant; c is the heat capacity of the gas and is given by a function of θ alone:

$$c(v,\theta) = \tilde{c}(\theta)\ . \tag{23}$$

(The relations (22) and (23) distinguish an ideal gas from other homogeneous fluid bodies.) Thus, for an ideal gas,

$$Q(\mathcal{P},\bar{L}) = \int_{\mathcal{L}_{\mathcal{P}}(\bar{L})} \left(\tilde{c}(\theta)d\theta + \frac{r\theta}{V}\,dV\right), \tag{24}$$

i.e.,

$$Q(\mathcal{P},\varphi^{-1}(\bar{\theta})) = \int_{\mathcal{L}_{\mathcal{P}}(\varphi^{-1}(\bar{\theta}))} \left(\tilde{c}(\theta)d\theta + \frac{r\theta}{V}\right)dV\ . \tag{25}$$

Serrin shows that his form of the second law implies that the functions φ corresponding to two distinct ideal gases must be proportional, and hence that ideal gases determine, to within a constant factor, a distinguished coordinate system on $\mathcal{H}$. In [1979,2] it is shown that the presence in $\mathbb{U}$ of elastic, or even viscous, substances far more general than ideal gases determines the same coordinatization of $\mathcal{H}$. This coordinatization, which is unique if, as in practice, one preassigns the value of the difference in the coordinates of the hotness levels of two phase transitions at a standard pressure, such as the freezing and boiling points of water at one atmosphere, is called the <u>absolute temperature scale</u>.

Serrin's principal result is that his statement of the second law is equivalent to asserting that <u>for every cycle</u> $\mathcal{P}$ <u>of each system in</u> $\mathbb{U}$

$$\int_0^\infty \frac{Q(\mathcal{P},\varphi^{-1}(\theta))}{\theta^2}\, d\theta \le 0 . \tag{26}$$

The importance of this relation, called the <u>accumulation inequality</u>, lies in its generality: it refers only to the absolute temperature scale φ and the "accumulation" $Q(\mathcal{P},\cdot)$ of the (countably additive) heat measure $q_{\mathcal{P}}$ on the hotness manifold $\mathcal{H}$; it is independent of the "space-time structure" or the concepts of "body" and "force" used in the specific physical theory to which the thermodynamical concepts of heat and hotness may be applied. In view of (19), when the function $\hat{Q}_{\mathcal{P}}$, defined by $\hat{Q}_{\mathcal{P}}(\theta) = Q(\mathcal{P},\varphi^{-1}(\theta))$, is of bounded variation, an integration by parts yields for sufficiently small $\delta > 0$,

$$\int_0^\infty \frac{Q(\mathcal{P},\varphi^{-1}(\theta))}{\theta^2}\, d\theta = \int_{\varphi(L^\ell)-\delta}^{\varphi(L^u)} \frac{\hat{Q}_{\mathcal{P}}(\theta)}{\theta^2}\, d\theta + \int_{\varphi(L^u)}^\infty \frac{\hat{Q}_{\mathcal{P}}(\varphi(L^u))}{\theta^2}\, d\theta$$

$$= \int_{\varphi(L^\ell)-\delta}^{\varphi(L^u)} \frac{d\hat{Q}_{\mathcal{P}}(\theta)}{\theta} - \frac{\hat{Q}_{\mathcal{P}}(\theta)}{\theta}\Bigg|_{\varphi(L^\ell)-\delta}^{\varphi(L^u)} + \frac{\hat{Q}_{\mathcal{P}}(\varphi(L^u))}{\varphi(L^u)}$$

$$= \int_{\varphi(L^\ell)-\delta}^{\varphi(L^u)} \frac{d\hat{Q}_{\mathcal{P}}(\theta)}{\theta} = \int_0^\infty \frac{d\hat{Q}_{\mathcal{P}}(\theta)}{\theta} . \tag{27}$$

Thus the accumulation inequality does give a mathematical form to the assertion that "the sum along a cycle of the ratio of heat gained to the absolute temperature at which it is gained cannot be positive."

The integral in the accumulation inequality plays the role of the action $\mathcal{S}$ in (2). It is clear that Serrin's form of the second law and his derivation of the accumulation inequality go a long way toward the resolution of problems (i), (ii), and (iii). In our recently published joint research with Serrin [1981,1], we have extended Serrin's form of the second law and the accumulation inequality so that they are meaningful for "approximate cycles". In this research, by combining definitions and methods of the papers [1974,1][1975,1],[1979,2,3], we answer the questions (i)-(iii) in a way that supplies not only identification of the action $\prec$ in (2) but also a derivation of the implication (3). We consider thermodynamical systems (Σ,Π,Q) that are systems (Σ,Π) in the sense explained above[10] and possess an accumulation function Q that assigns a number $Q(P,\sigma,L)$ to each triple (P,σ,L) with P in Π, σ in $\mathcal{D}(P)$, and L in the hotness manifold $\mathcal{H}$; $Q(P,\sigma,L)$ is called <u>the net heat absorbed by the system at levels of hotness at or below</u> L <u>in the process</u> P <u>starting at the state</u> σ. Let $\mathbb{P}$ be the set of pairs (P,σ) with P in Π and σ in $\mathcal{D}(P)$. In addition to a mild regularity condition for Q,[11] it is assumed that, for each (P,σ) in $\mathbb{P}$ [even if (P,σ) is not a cycle, i.e., does not have $\rho_P\sigma = \sigma$] there are hotness levels $L^\ell = L^\ell(P,\sigma)$ and $L^u = L^u(P,\sigma)$, with $L^\ell \prec L^u$, such that, in analogy with (19),

$$\left.\begin{aligned} Q(P,\sigma,L) &= 0 \quad \text{for } L \prec L^\ell , \\ Q(P,\sigma,L) &= Q(P,\sigma,L^u) \quad \text{for } L^u \prec L . \end{aligned}\right\} \tag{28}$$

We say that a pair (P,σ) in $\mathbb{P}$ is <u>absorptive</u> if $Q(P,\sigma,L) \geq 0$ for all L in $\mathcal{H}$; i.e., if the heat absorbed by the system at or below each level of hotness is not negative.[12]

The <u>union</u> $\mathcal{S}' \oplus \mathcal{S}''$ of two thermodynamical systems $\mathcal{S}' = (\Sigma',\Pi',Q')$ and $\mathcal{S}'' = (\Sigma'',\Pi'',Q'')$ is taken to be the system $\mathcal{S} = (\Sigma,\Pi,Q)$ with $\Sigma = \Sigma' \times \Sigma''$, $\Pi = \Pi' \times \Pi''$, $\mathbb{P} = \mathbb{P}' \times \mathbb{P}''$, and with

$$\rho_{(P',P'')}(\sigma',\sigma'') = (\rho_{P'}\sigma', \rho_{P''}\sigma'') \tag{29}$$

---

10) I.e., in the sense in which the word <u>system</u> is used in [1974,1] [1975,1]. It is observed in [1981,1] that many of the results given there hold under a concept of systems more general than that introduced in [1974,1][1975,1].

11) Namely that for each choice of P and σ the function $Q(P,\sigma,\cdot)$ is bounded and has at most a countable number of points of discontinuity.

12) When the processes correspond to functions of time, a pair (P,σ) may be absorptive and yet such that heat is emitted by the system during an interval of time; in such a case there will be (for the same process P and initial state σ) other intervals of time during which the system absorbs a compensating amount of heat.

and

$$Q((P',P''),(\sigma',\sigma''),L) = Q'(P',\sigma',L) + Q''(P'',\sigma'',L) , \qquad (30)$$

for each $((P',P''),(\sigma',\sigma''))$ in $\mathbb{P} = \mathbb{P}'\times \mathbb{P}''$ and each L in $\mathcal{H}$. As in [1979,2,3], it is assumed that a collection $\mathbf{U}$ of thermodynamical systems, closed under the union operation, is given and that $\mathbf{U}$ contains at least one special system that corresponds to an elastic or viscous substance. Again, the discussion of the second law takes its simplest form if the distinguished systems are ideal gases. Each state σ of an ideal gas is represented as a point (V,θ) in the first quadrant of a coordinate plane; each process P_t of the gas is a piecewise continuous function on an interval $[0,t)$ with values $(\dot{V}(\tau), \dot{\theta}(\tau))$ that, for each τ in $[0,t)$, can be interpreted as the rates of change of V and θ at time τ ; a pair (P_t,σ°) , with $\sigma^\circ = (V^\circ,\theta^\circ)$, is in $\mathbb{P}$ if, for each s in $[0,t]$, $(V(s),\theta(s))$, with

$$\left.\begin{aligned} V(s) &= V^\circ + \int_0^s \dot{V}(\tau)d\tau , \\ \theta(s) &= \theta^\circ + \int_0^s \dot{\theta}(\tau)d\tau , \end{aligned}\right\} \qquad (31)$$

is in Σ , i.e., has $V(s) > 0$ and $\theta(s) > 0$; in such a case

$$\rho_{P_t}\sigma^\circ = (V(t),\theta(t)) . \qquad (32)$$

It is again part of the definition of an ideal gas that θ is given by a coordinate system φ on $\mathcal{H}$ and no more is assumed about φ than that it is a strictly increasing, positive-valued, continuous function. For an ideal gas the function Q has the form

$$Q(P,\sigma^\circ,\bar{L}) = \int_{M(P_t,\sigma,\bar{L})} \left(\tilde{c}(\theta(s))\dot{\theta}(s) + \frac{r\theta(s)}{V(s)}\dot{V}(s)\right) ds, \qquad (33)$$

where r is a number, $\tilde{c}$ is a function characteristic of the gas, and

$$M(P_t,\sigma,\bar{L}) = \{s \mid 0 \leq s < t,\ \theta(s) \leq \varphi(\bar{L}) . \qquad (34)$$

When the curve with the parameterization (31) on $[0,t]$ is a closed curve, and hence $V(t) = V^\circ$, $\theta(t) = \theta^\circ$, (32) yields $\rho_{P_t}\sigma^\circ = \sigma^\circ$, and the pair (P_t,σ°) is a cycle; in such a case, if we write ρ for

$(P_t,\sigma°)$, the equations (33) and (24) become the same. The equation (33), which can be written in the line-integral notation used in equations (24) and (25), is an extension of these equations from pairs $(P_t,\sigma°)$ that are cycles to pairs $(P_t,\sigma°)$ with $\sigma°$ in $\mathfrak{D}(P_t)$, i.e., from $\mathbb{P}_c$ to $\mathbb{P}$.

We take the second law to be the following statement that pertains to each system $\mathcal{S} = (\Sigma,\Pi,Q)$ in $\mathbb{U}$: <u>For each level</u> $\bar{L}$ <u>of hotness and each</u> $\varepsilon > 0$, <u>each state</u> σ <u>has a neighborhood</u> $\mathfrak{G}_\varepsilon(\sigma,\bar{L})$ <u>in</u> Σ <u>for which</u>

$$\rho_P\sigma \in \mathfrak{G}_\varepsilon(\sigma,\bar{L}) \ , \quad (P,\sigma) \ \text{absorptive}, \ L^u(P,\sigma) \prec \bar{L}$$

$$\Rightarrow 0 \leq Q(P,\sigma,L^u(P,\sigma)) < \varepsilon \ . \tag{35}$$

In terms more suggestive but less precise: the overall net gain of heat is small in an approximate cycle that is absorptive and operates at or below a fixed level $\bar{L}$ of hotness.

In (35), the relation $\rho_P\sigma \in \mathfrak{G}_\varepsilon(\sigma,\bar{L})$ indicates an "approximate cycle" and the relation $L^u = L^u(P,\sigma) \prec \bar{L}$ is the assertion that a pair (P,σ) "operates at or below the fixed level $\bar{L}$ of hotness". The relation $0 \leq Q(P,\sigma,L^u)$ is true for any absorptive pair (P,σ) . The relation $Q(P,\sigma,L^u) < \varepsilon$, however, is the assertion that the "overall net gain of heat is small" and is the important conclusion of the implication (35).

It is a consequence of this law that the hotness manifold again has a distinguished coordinate system φ that is unique to within a constant multiple, and this coordinate system is that employed in the formula (33) for the accumulation function of an ideal gas. The principal results obtained in [1981,1] are of the following type: <u>The second law is equivalent to the assertion that for every</u> $\bar{L}$ <u>in</u> $\mathcal{H}$ <u>and every thermodynamical system</u> (Σ,Π,Q) <u>in</u> $\mathbb{U}$, <u>each state</u> $\sigma°$ <u>has, for each</u> $\varepsilon > 0$, <u>a neighborhood</u> $\mathfrak{G}_\varepsilon(\sigma°,\bar{L})$ <u>in</u> Σ <u>for which</u>

$$\int_0^\infty \frac{Q(P,\sigma°,\varphi^{-1}(\theta))}{\theta^2}\, d\theta < \varepsilon \tag{36}$$

<u>whenever</u> $(P,\sigma°)$ <u>is in</u> $\mathbb{P}$, $L^u(P,\sigma) \prec L$, <u>and</u>

$$\rho_P\sigma \in \mathfrak{G}_\varepsilon(\sigma,\bar{L}) \ . \tag{37}$$

In other words: the second law holds if and only if each system in

$\mathbb{U}$ is such that its accumulation integral is approximately negative on approximate cycles. In particular, the second law implies that when

$$\mathcal{S}(P,\sigma^\circ) = \int_0^\infty \frac{Q(P,\sigma^\circ,\varphi^{-1}(\theta))}{\theta^2}\, d\theta$$

and (P,σ°) operates at or below $\bar{L}$, the implication (3), with $\mathfrak{G}_\varepsilon(\sigma^\circ) = \mathfrak{G}_\varepsilon(\sigma^\circ,\bar{L})$, holds for each system in $\mathbb{U}$, each state σ°, and each $\varepsilon > 0$.

In the second part of this essay we have discussed the problem of characterizing the action $\mathcal{A}$ in a general, essentially context-free, manner in which only the concepts of heat and hotness need be mentioned. Of course, in the thermomechanics of continuous media, formulae for $\mathcal{S}$ have long been known. Of primary interest to researchers in that field will be the ideas and theorems presented in the first part of our discussion, namely in the paragraphs containing the relations (1)-(18). The concepts set forth there give us an approach to thermodynamics in which the existence and regularity of entropy (and of free energy) as a function of state is to be deduced rather than assumed.[13] In our first paper employing this new approach [1974,1], we examined the problem of finding the restrictions that the second law places on the constitutive equations of elastic and viscous materials, materials with internal state variables, and materials with fading memory, and we found that the assumption that $\mathcal{S}$ has the Clausius property yields restrictions on the response functions (or functionals) that give such experimentally observable quantities as stress, heat flux, and internal energy (or temperature) agreeing perfectly with restrictions obtained in the treatments that start with a differentiable entropy, or free energy, function and employ the Clausius-Duhem inequality [1963,1,2]

---

[13] The need for such an approach was brought out in the book of Day [1972,2] which predated and influenced our paper of 1974. (Another important influence was a paper of Noll [1972,3] which showed the usefulness of the concept of state for a broad class of materials, including materials with memory.) Day, starting with a Clausius-type inequality for non-linear materials with fading memory, was able to obtain, albeit under neglect of the contribution to $\mathcal{S}$ of a term involving the inner product of the heat flux vector and the temperature gradient, the existence of entropy and free energy functionals; however, he assumed, rather than proved, that these functionals have the smoothness needed to proceed further and derive Coleman's formula for stress as an instantaneous derivative of free energy. Only the equilibrium response was shown to have the smoothness expected of it. Nevertheless, Day's book stimulated the search for general Clausius- and Kelvin-type formulations of the second law by showing that, even for materials with memory, the existence of entropy can be proved from such starting points.

[1964,1-3][1967,1,2]. For each of these materials more was known about $\mathcal{s}(P,\sigma)$ than its continuity in σ or its general representation as an accumulation integral, and consequently more was proven about entropy and free energy than semi-continuity. In each case it was shown that, *starting with an appropriate expression for* $\mathcal{s}$ *in terms of the response functions or functionals for stress, heat flux, and internal energy, and assuming that* $\mathcal{s}$ *has the Clausius property, one can construct an entropy function (or functional) with the properties of differentiability needed to derive the principal results of the earlier studies*. The failure of the entropy and free energy of a material with fading memory to be unique does not invalidate the earlier studies based on the Clausius-Duhem inequality. The implications of the earlier work for the response functionals for stress, heat flux, and internal energy, required only the existence of an appropriately smooth free-energy functional, not its uniqueness.

It has been found that for some materials one can separate the problem of finding the class of entropy functions from that of deriving the thermodynamical restrictions on response functions relating experimentally accessible quantities. For example, the local thermomechanics of a unidimensional elastic-plastic material (with its elastic behavior linear and its plastic behavior perfect) is described by giving the elastic modulus μ and the yield strain α as functions of the temperature θ, and the heat capacity κ, the latent elastic heat Λ_e and the latent plastic heat Λ_p as functions of the elastic strain λ_e, the plastic strain λ_p, and the temperature:

$$\mu = \mu(\theta) \ , \quad \alpha = \alpha(\theta) \ ,$$

$$\kappa = \kappa(\lambda_e,\lambda_p,\theta), \quad \Lambda_e = \Lambda_e(\lambda_e,\lambda_p,\theta), \quad \Lambda_p = \Lambda_p(\lambda_e,\lambda_p,\theta) \ .$$

Without mention of entropy or free energy, one can derive relations among the functions $\mu,\sigma,\kappa,\Lambda_e,\Lambda_p$ that are necessary and sufficient for compliance with the laws of thermodynamics (see [1976,2],[1979,1]). One may separately find conditions on these functions sufficient for the entropy function, normalized as in equation (11), to be unique and, in cases where entropy and free energy are not unique, the class of such functions can be precisely described [1975,1][1979,1].[14)]

We are grateful to James Serrin for the opportunity to work with

---

[14)] We discuss corresponding problems for hypoelastic materials in [1976,1] and there show that each hypoelastic material has a unique normalized free energy function.

him on the theory of the accumulation inequality.

The preparation of this essay was supported in part by the U.S. National Science Foundation and the Italian National Council for Research.

References

1896 [1] Mach, E.: Die Prinzipien der Wärmelehre, Historisch-kritisch entwickelt. Leipzig, Barth.

1963 [1] Coleman, B. D., & V. J. Mizel: Thermodynamics and departures from Fourier's law of heat conduction. Arch. Rational Mech. Anal. **13**, 245-261.

[2] Coleman, B. D., & W. Noll: The thermodynamics of elastic materials with heat conduction and viscosity. Arch. Rational Mech. Anal. **13**, 167-178.

1964 [1] Coleman, B. D.: Thermodynamics of materials with memory, Arch. Rational Mech. Anal. **17**, 1-46.

[2] Coleman, B. D.: Thermodynamics, strain impulses, and visco-elasticity. Arch. Rational Mech. Anal. **17**, 230-254.

[3] Coleman, B. D., & V. J. Mizel: Existence of caloric equations of states in thermodynamics. J. Chem. Phys. **40**, 1116-1125.

1967 [1] Coleman, B. D., & M. E. Gurtin: Equipresence and constitutive equations for rigid heat conductors. Z.A.M.P. **18**, 199-208.

[2] Coleman, B. D., & M. E. Gurtin: Thermodynamics with internal state variables. J. Chem. Phys. **47**, 597-613.

1972 [1] Boyling, J. B.: An axiomatic approach to classical thermodynamics. Proc. R. Soc. London **A 329**, 35-70.

[2] Day, W. A.: The Thermodynamics of Simple Materials with Fading Memory. Springer Tracts in Natural Philosophy, Vol. 22. Berlin, etc.: Springer.

[3] Noll, W.: A new mathematical theory of simple materials. Arch. Rational Mech. Anal. **48**, 1-50.

1974 [1] Coleman, B. D., & D. R. Owen: A mathematical foundation for thermodynamics. Arch. Rational Mech. Anal. **54**, 1-104.

1975 [1] Coleman, B. D., & D. R. Owen: On thermodynamics and elastic-plastic materials. Arch. Rational Mech. Anal. **59**, 25-51.

1976 [1] Coleman, B. D., & D. R. Owen On thermodynamics and intrinsically equilibrated materials. Annali Mat. pura applicata (IV) **108**, 189-199.

[2] Coleman, B. D., & D. R. Owen: Thermodynamics of elastic-plastic materials. Rend. Accad. Naz. Lincei (VIII) Classe di Scienze, fis. mat. nat. **61**, 77-81.

1977 [1] Truesdell, C., & S. Bharatha: The Concepts and Logic of Classical Thermodynamics as a Theory of Heat Engines, Rigorously Developed upon the Foundation Laid by S. Carnot and R. Reech, New York, etc.: Springer.

1979 [1] Coleman, B. D., & D. R. Owen: On the thermodynamics of elastic-plastic materials with temperature-dependent moduli and yield stresses. Arch. Rational Mech. Anal. **70**, 339-354.

[2] Serrin, J.: Lectures on Thermodynamics, University of Naples.

[3] Serrin, J.: Conceptual analysis of the classical second laws of thermodynamics. Arch. Rational Mech. Anal. **70**, 355-371.

[4] Truesdell, C.: Absolute temperature as a consequence of Carnot's General Axiom. Arch. History Exact Sci. **20**, 357-380.

1980 [1] Šilhavý, M.: On measures, convex cones, and foundations of thermodynamics; I. Systems with vector-valued actions; II. Thermodynamic Systems. Czech. J. Phys. **B 30**, 841-861, 961-991.

[2] Truesdell, C.: The Tragicomical History of Thermodynamics 1822-1854, New York, etc.: Springer.

1981 [1] Coleman, B. D., D. R. Owen, & J. Serrin: The second law of thermodynamics for systems with approximate cycles, Arch. Rational Mech. Anal. **77**, 103-142.

1982 [1] Feinberg, M., & R. Lavine: Thermodynamics based on the Hahn-Banach Theorem: the Clausius inequality, Arch. Rational Mech. Anal., **82**, 202 - 293.

[2] Owen, D. R.: The second law of thermodynamics for semi-systems with few approximate cycles, Arch. Rational Mech. Anal. **80**, 39-55.

[3] Silhavy, M.: On the Clausius inequality, Arch. Rational Mech. Anal., **81**, 221 - 243.

GLOBAL AND LOCAL VERSIONS OF THE SECOND LAW OF THERMODYNAMICS

David R. Owen

Department of Mathematics
Carnegie-Mellon University
Pittsburgh, Pennsylvania 15213

1. Introduction.

Two important directions of recent research on the mathematical foundations of thermodynamics have led to what I call evolutionary and geometrical theories of thermodynamics. In the former, emphasis is laid on the states and processes of a physical system and on interactions between the system and its environment which accompany changes of state. In the latter, a system is required to have the structure of a continuous, three-dimensional body which, together with its collection of subbodies, is considered at a fixed instant in time in order to emphasize geometrical aspects of continuum thermodynamics. Evolutionary theories are algebraic and topological in nature and have helped to justify the use of functions of state such as energy and entropy for a broad class of physical systems which includes as a proper subclass the systems studied in classical thermodynamics. Geometrical theories are largely measure-theoretical in nature and have clarified the relation between global and local statements of the first and second laws of thermodynamics by providing conditions under which global and local versions of these laws are mathematically equivalent. Much of the modern research on evolutionary aspects of thermodynamics is contained in papers by DAY [1969], NOLL [1972], and COLEMAN & OWEN [1974,1977], while many of the basic concepts and results of the geometrical approach are to be found in the work of NOLL [1959,1963], GURTIN & WILLIAMS [1967], and WILLIAMS [1970].

In this paper I present some preliminary work on the problem of unifying evolutionary and geometrical theories of thermodynamics. Although earlier work by WILMANSKI [1972] and WILLIAMS [1974] has provided promising approaches to this problem, there is lacking at present a body of concepts and results which encompasses the major content of the two component theories. It is my hope that the research reported on here can serve as a guide for more complete and more general unifications that would include not only algebraic aspects of evolutionary theories and measure-theoretical aspects of geometrical theories, as are included here, but also topological aspects of evolutionary theories. The exclusion of topological aspects made here removes from within the scope of the present unified theory those continuous bodies comprised of a material with fading memory. Although this is a serious defect, the theory outlined below gains in clarity and simplicity from the exclusion, particularly when it comes to discussions of the important and difficult concept of "local state". Other shortcomings of the present approach are the exclusions of any distinction between volume and surface interactions and of "unbalanced" interactions. Removal of these defects would involve geometrical ideas closer to those proposed by WILLIAMS [1970] and would greatly complicate the presentation here.

Among the various ideas I have attempted to incorporate in a unified theory of thermodynamics, three deserve separate mention in this introduction. The first is the idea that the states and processes of a subbody are accessible only via states and processes of the entire body. This amounts to the assertion that a subbody of a continuous body cannot change its state without entailing a change of state of the larger body of which it is a part. This attribute of a continuous body may be called its <u>integrity</u>. (One aspect of the integrity of a continuous body is seen in NOLL's description of a body as a differentiable, three-dimensional manifold: there is an atlas for this manifold consisting entirely of globally defined charts.)

The second idea which I wish to emphasize is that the notion of a material point should be a derived one in any fundamental treatment of continuum thermodynamics. It is easy for one familiar with the field theories of continuum physics to accept the notion of a material point as a primitive concept, because it leads to a simple and useful means of distinguishing one material from another, namely, through the "local equations of state" or the "constitutive relation" of a material. Here I take as basic the fact that a material is accessible to us only in three-dimensional pieces. Any concept of material point, no matter how useful, should be consistent with this fact, in the sense that the states and processes which a material point can experience are obtained in a specified manner from the states and processes of any subbody whose interior contains that material point. In my opinion, it is only through a consistency requirement of this type that the concept of a material point can meet the standards of clarity and rigor associated with modern studies of the foundations of continuum physics.

The third and final idea which deserves emphasis here is a recognition of the relative paucity of cyclic (or even "approximately cyclic") processes of a continuous body. A corresponding situation has been emphasized in the work of COLEMAN and OWEN on material elements with fading memory [1974, Section 12] and on semi-systems with restrictions on the accessibility of states [1977]. That research motivates the version of the second law used here. In this version, no mention is made of the notion of a cyclic process.

As a final introductory remark, I wish to mention that the basic mathematical concept employed in the present theory is that of a "pre-sheaf of algebraic semi-systems." An "algebraic semi-system" is similar to the concept of "semi-system" studied by COLEMAN and OWEN [1977], but their requirement of a topological structure on the state space is not made here, so that only the algebraic features of their concept are retained here. A "pre-sheaf of algebraic semi-systems" is formed by assigning to each part of a continuous body an algebraic semi-system in such a way that processes and states of larger subbodies determine processes and states of smaller subbodies in a manner consistent with set inclusion. The natural appearance here of a concept from sheaf theory may provide a further point of contact between continuum physics and category theory.

2. Basic evolutionary and global geometrical concepts

Our first two definitions concern the "evolutionary" aspects of physical systems.

Definition 2.1. An algebraic semi-system is a pair (Σ,Π) of sets, with the elements σ of Σ called states and the elements P of Π called processes, together with two functions

$$P \mapsto \rho_P \tag{2.1}$$

$$(P'',P') \mapsto P''P' . \tag{2.2}$$

The function $P \mapsto \rho_P$ assigns to each process P a function $\rho_P : \mathcal{D}(P) \to \Sigma$, called the transformation induced by P , with $\mathcal{D}(P)$ a non-empty subset of Σ , and satisfies the following condition:

(I) there exists a state σ in Σ such that

$$\Pi\sigma := \{\rho_P\sigma \mid P \in \Pi,\ \sigma \in \mathcal{D}(P)\} \qquad (2.3)$$

equals Σ .

The function $(P'',P') \mapsto P''P'$ maps the set

$$\mathbb{P} := \{(P'',P') \in \Pi \times \Pi \mid \rho_{P'}^{-1}(\mathcal{D}(P'')) \neq \emptyset\} \qquad (2.4)$$

into Π and is such that

(II) for each pair (P'',P') in $\mathbb{P}$, the transformation $\rho_{P''P'}$ induced by the process $P''P'$ has its domain $\mathcal{D}(P''P')$ given by

$$\mathcal{D}(P''P') = \rho_{P'}^{-1}(\mathcal{D}(P'')) , \qquad (2.5)$$

and, for each σ in $\mathcal{D}(P''P')$, there holds

$$\rho_{P''P'}\sigma = \rho_{P''}\rho_{P'}\sigma .$$

The process $P''P'$ is called the <u>successive application of</u> P' <u>and</u> P'' , or P' <u>followed by</u> P'' .

We write $\Pi \lozenge \Sigma$ for the set of pairs $\{(P,\sigma) \mid \sigma \in \Sigma,\ P \in \Pi$, and $\sigma \in \mathcal{D}(P)\}$, and we often write $P\sigma$ for the state $\rho_P\sigma$, called the <u>final state</u> associated with the pair (P,σ) in $\Pi \lozenge \Sigma$. For each state σ , the set $\Pi\sigma$ defined in (2.3) is called the <u>set</u> of states accessible from σ , and (I) asserts that there is a state from which all states in Σ are accessible. Such a state is called a <u>base state</u> of (Σ,Π) , so that (I) amounts to the condition that (Σ,Π) have at least one base state. (An algebraic semi-system <u>all</u> of whose states are base states has been called elsewhere a "system with perfect accessibility".)

<u>Definition 2.2</u>. An <u>action</u> $\mathcal{a}$ for an algebraic semi-system (Σ,Π) is a real-valued function on $\Pi \lozenge \Sigma$ which is additive with respect to successive application, i.e., for each pair (P'',P') in $\mathbb{P}$ and state σ in $\mathcal{D}(P''P')$, there holds

$$\mathcal{a}(P''P',\sigma) = \mathcal{a}(P',\sigma) + \mathcal{a}(P'',P'\sigma) . \qquad (2.6)$$

The number $\mathcal{a}(P,\sigma)$ is called the <u>supply of</u> $\mathcal{a}$ when the semi-system undergoes the process P starting at the state σ , and (2.6) means that the supply of $\mathfrak{a}$ when the semi-system undergoes $P''P'$ starting at σ is the sum of the supply of $\mathfrak{a}$ when the semi-system undergoes P' starting at σ and the supply of $\mathfrak{a}$ when the semi-system undergoes P'' starting at the final state associated with (P',σ) .

The next definition permits us to relate two algebraic semi-systems by means of mappings which preserve their structure.

<u>Definition 2.3</u>. A <u>morphism</u> Ψ of (Σ,Π) into $(\hat{\Sigma},\hat{\Pi})$ is a function from $\Sigma \cup \Pi$ into $\hat{\Sigma} \cup \hat{\Pi}$ such that

(M1) $\Psi(\Sigma) \subset \hat{\Sigma}$, $\Psi(\Pi) \subset \hat{\Pi}$;

(M2) for all P in Π ,

$$\mathfrak{D}(\Psi P) = \Psi(\mathfrak{D}(P)) \text{ ,} \tag{2.7}$$

and, for all (P,σ) in $\Pi \diamond \Sigma$,

$$\Psi(P\sigma) = \Psi P \Psi \sigma \text{ ;} \tag{2.8}$$

(M3) for all (P'',P') in $\mathbb{P}$, there holds

$$\Psi(P''P') = \Psi P'' \Psi P' \text{ .} \tag{2.9}$$

Relation (2.7) implies that $\Psi(\Pi \diamond \Sigma)$ is a subset of $\hat{\Pi} \diamond \hat{\Sigma}$, so that the right hand side of (2.8) is meaningful. If $\mathfrak{R}(P)$ denotes the range of ρ_P , then (2.7) and (2.8) imply that

$$\mathfrak{R}(\Psi P) = \Psi \mathfrak{R}(P) \text{ .} \tag{2.10}$$

For each (P'',P') in $\mathbb{P}$, there then holds

$$\begin{aligned} \Psi(\mathfrak{D}(P'') \cap \mathfrak{R}(P')) &\subset \Psi(\mathfrak{D}(P'')) \cap \Psi(\mathfrak{R}(P')) \\ &= \mathfrak{D}(\Psi P'') \cap \mathfrak{R}(\Psi P') \text{ ;} \end{aligned}$$

hence, $\mathfrak{D}(\Psi P'') \cap \mathfrak{R}(\Psi P')$ is non-empty. In other words, $\rho^{-1}_{\Psi P'}(\mathfrak{D}(\Psi P''))$ is non-empty whenever (P'',P') is in $\mathbb{P}$, and (2.4) yields $\Psi(\mathbb{P}) \subset \hat{\mathbb{P}}$. Thus, the right-hand side of (2.9) is meaningful whenever the left-hand side is.

The previous discussion yields the relations

$$\Psi(\Sigma) \subset \hat{\Sigma} \tag{2.11}$$

$$\Psi(\Pi) \subset \hat{\Pi} \tag{2.12}$$

$$\Psi(\Pi \lozenge \Sigma) \subset \hat{\Pi} \lozenge \hat{\Sigma} \tag{2.13}$$

$$\Psi(\mathbb{P}) \subset \hat{\mathbb{P}} \tag{2.14}$$

for a morphism Ψ of an algebraic semi-system (Σ,Π) into a second algebraic semi-system $(\hat{\Sigma},\hat{\Pi})$. In addition, if equality holds in (2.11), then

$$\Psi(\Sigma_{base}) \subset \hat{\Sigma}_{base}, \tag{2.15}$$

where Σ_{base} and $\hat{\Sigma}_{base}$ denote the collections of base states for (Σ,Π) and $(\hat{\Sigma},\hat{\Pi})$, respectively. The relations (2.11)-(2.15) indicate that there may be some freedom when it comes to defining the notion of an epimorphism of algebraic semi-systems: one might choose from among (2.11)-(2.15) certain relations and require in these that "inclusion" becomes "equality". It turns out that the appropriate definition of epimorphism for present purposes results from requiring equality in relations (2.11)-(2.14) but not in (2.15). Thus, an <u>epimorphism</u> is defined to be a morphism Ψ which satisfies

$$\left.\begin{array}{ll} \Psi(\Sigma) = \hat{\Sigma} & \Psi(\Pi) = \hat{\Pi} \\ \Psi(\Pi \lozenge \Sigma) = \hat{\Pi} \lozenge \hat{\Sigma}, & \Psi(\mathbb{P}) = \hat{\mathbb{P}} \end{array}\right\} . \tag{2.16}$$

We introduce geometrical concepts now by considering a bounded, regularly closed subset $\mathcal{B}$ of physical space. (A regularly closed set equals the closure of its interior.) An <u>algebra of subbodies</u> (or <u>parts</u>) of $\mathcal{B}$ is a collection $\mathbb{B}$ of regularly closed subsets $\mathcal{P}$ of $\mathcal{B}$ which contains $\mathcal{B}$, the relative exterior $cl(\mathcal{B}\setminus\mathcal{P})$ of each part $\mathcal{P}$, the (regularly closed) overlap of $\mathcal{B}$ and any solid box, and which is closed under the operations $\vee,\wedge$:

$$\mathcal{P} \vee \mathcal{Q} = \mathcal{P} \cup \mathcal{Q} \quad \text{and} \quad \mathcal{P} \wedge \mathcal{Q} = cl(\text{int}\,\mathcal{P} \cap \text{int}\,\mathcal{Q}) .$$

(Closure under $\wedge$ follows from the other conditions on $\mathbb{B}$.) Each element $\mathcal{P}$ of $\mathbb{B}$ is called a <u>subbody</u>, or <u>part</u>, of $\mathcal{B}$, while $\mathcal{B}$ is called the <u>body</u> for the collection $\mathbb{B}$. Additional requirements on $\mathbb{B}$ are made in the references on geometrical theories of thermodynamics mentioned in the introduction, but these requirements are not needed for the results obtained here.

The next definition combines evolutionary and geometrical ideas to form the main concept of the present theory.

Definition 2.4. Let $\mathbb{B}$ be an algebra of subbodies of $\mathcal{B}$. A pre-sheaf of algebraic semi-systems over $\mathcal{B}$ is a family $((\Sigma_{\mathcal{P}},\Pi_{\mathcal{P}})\,|\,\mathcal{P} \in \mathbb{B})$ of algebraic semi-systems together with a family $(\Psi^{\mathcal{Q}}_{\mathcal{P}}\,|\,\mathcal{P} \subset \mathcal{Q},\ \mathcal{P},\mathcal{Q} \in \mathbb{B})$ of mappings satisfying the following conditions:

(PS1) for each $\mathcal{P},\mathcal{Q} \in \mathbb{B}$, with $\mathcal{P} \subset \mathcal{Q}$, the mapping $\Psi^{\mathcal{Q}}_{\mathcal{P}}$ is an epimorphism of $(\Sigma_{\mathcal{Q}},\Pi_{\mathcal{Q}})$ onto $(\Sigma_{\mathcal{P}},\Pi_{\mathcal{P}})$;

(PS2) for each $\mathcal{P} \in \mathbb{B}$, the mapping $\Psi^{\mathcal{P}}_{\mathcal{P}}$ is the identity morphism on $(\Sigma_{\mathcal{P}},\Pi_{\mathcal{P}})$;

(PS3) for each $\mathcal{P},\mathcal{Q},\mathcal{R} \in \mathbb{B}$ with $\mathcal{P} \subset \mathcal{Q} \subset \mathcal{R}$ there holds

$$\Psi^{\mathcal{R}}_{\mathcal{P}} = \Psi^{\mathcal{Q}}_{\mathcal{P}} \circ \Psi^{\mathcal{R}}_{\mathcal{Q}} . \tag{2.17}$$

This definition embodies both the geometrical structure of a continuous body and the ability of each part of a body to evolve, i.e., to change its state. Moreover, it provides connections between the evolution of the various parts of a body. For example, (PS1) requires that when a part $\mathcal{Q}$ include a part $\mathcal{P}$, then every pair $(P_{\mathcal{P}},\sigma_{\mathcal{P}})$ in $\Pi_{\mathcal{P}} \lozenge \Sigma_{\mathcal{P}}$ is induced by a pair in $\Pi_{\mathcal{Q}} \lozenge \Sigma_{\mathcal{Q}}$. In other words, any evolution of $\mathcal{P}$ entails an evolution of $\mathcal{Q}$. In particular, since every $\mathcal{P}$ in $\mathbb{B}$ is a subset of the body $\mathcal{B}$, every evolution of a part of $\mathcal{B}$ is associated with an evolution of $\mathcal{B}$ itself. (This is the notion of integrity of a body discussed in the introduction.) Condition (PS3) is a consistency condition which assures that intermediate parts lying between two given comparable parts evolve in a way which is compatible with the evolution of the given parts.

The implications of the particular choice of the concept of epimorphism made above now can be stated more clearly. Our choice affords a strong notion of integrity for a body while permitting a greater degree of accessibility of states for a smaller part than for a larger one. For example, it may happen that $\mathcal{P} \subset \mathcal{Q}$ and

$$\Psi^{\mathcal{Q}}_{\mathcal{P}}((\Sigma_{\mathcal{Q}})_{\mathrm{base}}) \subsetneq (\Sigma_{\mathcal{P}})_{\mathrm{base}} ,$$

i.e., not every base state for $\mathcal{P}$ is induced by a base state of $\mathcal{Q}$. Thus, in the terminology mentioned earlier, it is conceivable that $(\Sigma_{\mathcal{P}},\Pi_{\mathcal{P}})$ can have many base states and even be a system with perfect accessibility, whereas $(\Sigma_{\mathcal{Q}},\Pi_{\mathcal{Q}})$ has only one base state. Such possibilities must not be excluded, particularly for the "local" algebraic semi-systems constructed in the next section, for it is often the case that material points can be described by systems with perfect accessibility while the parts of a body cannot.

3. Material points.

Let $\mathbb{B}$ be an algebra of subbodies of a body $\mathcal{B}$, let x be in the interior of $\mathcal{B}$, and let $\mathbb{B}_x$ denote the parts of $\mathcal{B}$ whose interiors contain x. For a given pre-sheaf over $\mathcal{B}$ we make the following definition.

Definition 3.1. Let $\mathcal{P},\mathcal{Q}$ be in $\mathbb{B}_x$ and let $\sigma_{\mathcal{P}} \in \Sigma_{\mathcal{P}}$, $\sigma_{\mathcal{Q}} \in \Sigma_{\mathcal{Q}}$. We say that $\sigma_{\mathcal{P}}$ and $\sigma_{\mathcal{Q}}$ are equivalent at x, and write $\sigma_{\mathcal{P}} \sim_x \sigma_{\mathcal{Q}}$, if there exists a part $\mathcal{R}$ contained in $\mathcal{P} \wedge \mathcal{Q}$ such that $\mathcal{R}$ is in $\mathbb{B}_x$ and

$$\Psi^{\mathcal{P}}_{\mathcal{R}}\sigma_{\mathcal{P}} = \Psi^{\mathcal{Q}}_{\mathcal{R}}\sigma_{\mathcal{Q}} . \tag{3.1}$$

Processes $P_{\mathcal{P}}$ and $P_{\mathcal{Q}}$ are said to be equivalent at x ($P_{\mathcal{P}} \sim_x P_{\mathcal{Q}}$) if a similar condition holds with states replaced by processes.

This notion of equivalence is a strong one, because it requires that a subbody be found on which both states (or both processes) have the same projection. For states, $\sim_x$ is a relation on $\bigcup_{\mathcal{P}\in\mathbb{B}_x} \Sigma_{\mathcal{P}}$ and for process on $\bigcup_{\mathcal{P}\in\mathbb{B}_x} \Pi_{\mathcal{P}}$, and in both cases $\sim_x$ is an equivalence relation. The strength of $\sim_x$ is reflected in our expectation that there be many equivalence classes $\sigma_x \subset \bigcup_{\mathcal{P}\in\mathbb{B}_x} \Sigma_{\mathcal{P}}$ and $P_x \subset \bigcup_{\mathcal{P}\in\mathbb{B}_x} \Pi_{\mathcal{P}}$. It is natural to write Σ_x for the collection of equivalence classes σ_x and Π_x for the collection of equivalence classes P_x, and to call $\sigma_x[P_x]$ a local state [process] of $\mathcal{B}$ at x. For each $\mathcal{P} \in \mathbb{B}_x$ the mapping of $\Sigma_{\mathcal{P}} \cup \Pi_{\mathcal{P}}$ into $\Sigma_x \cup \Pi_x$ which assigns to states and processes of $\mathcal{P}$ the local states and local processes to which they belong is denoted by $\Psi^{\mathcal{P}}_x$. It is easy to show that $\Psi^{\mathcal{P}}_x$ is a surjection, and there holds

$$\mathcal{P} \subset \mathcal{Q} \Rightarrow \Psi_x^{\mathcal{Q}} = \Psi_x^{\mathcal{P}} \circ \Psi_{\mathcal{P}}^{\mathcal{Q}} . \qquad (3.2)$$

In particular, the projections $\{\Psi_x^{\mathcal{Q}} | \mathcal{Q} \in \mathbb{B}_x\}$ as well as the pair (Σ_x, Π_x) are determined by $(\Sigma_{\mathcal{P}}, \Pi_{\mathcal{P}})$ for any one element $\mathcal{P}$ of $\mathbb{B}_x$, so it is convenient to deal only with the projection $\Psi_x^{\mathcal{B}}$ of $\Sigma_{\mathcal{B}} \cup \Pi_{\mathcal{B}}$ onto $\Sigma_x \cup \Pi_x$.

The pair (Σ_x, Π_x) can be given the structure of an algebraic semi-system in a natural way which makes the projection $\Psi_x^{\mathcal{B}}$ an epimorphism of algebraic semi-systems. [For example, one can define for each $P_x \in \Pi_x$,

$$\mathcal{D}(P_x) = \Psi_x^{\mathcal{B}} \mathcal{D}(P_{\mathcal{B}}) , \text{ with } P_{\mathcal{B}} \in P_x$$

and

$$P_x \sigma_x = \Psi_x^{\mathcal{B}}(P_{\mathcal{B}} \sigma_{\mathcal{B}}) , \quad \sigma_{\mathcal{B}} \in \mathcal{D}(P_{\mathcal{B}}) \cap \sigma_x , \quad P_{\mathcal{B}} \in P_x .$$

These definitions are meaningful and determine the set of compatible pairs $\Pi_x \lozenge \Sigma_x$.] It is appropriate to call the algebraic semi-system (Σ_x, Π_x) derived from the given pre-sheaf the <u>material point at</u> x. Thus, we have an explicit way of obtaining the structure associated with a point in a continuous body as it arises from the structure prescribed on all of its parts.

4. <u>Global and local actions</u>.

Although our notion of integrity of a body requires that the states and processes of larger parts determine those of smaller parts, the same need not be true for the interactions between a part and its exterior. For example, knowledge of the amount of heat which a body gains from its exterior does not always tell us how much heat any proper subbody gains from its exterior. Therefore, we need to prescribe such interactions for every part of a body. Moreover, in order to localize these interactions at a point, we must be able to compute densities of interactions. These considerations lead us to the next definition. In this definition and throughout the rest of this paper, there is given a pre-sheaf of algebraic semi-systems over a body $\mathcal{B}$.

<u>Definition 4.1</u>. A global action family $(\mathfrak{a}_{\mathcal{P}} | \mathcal{P} \in \mathbb{B})$ is a family in

which, for each $\mathcal{P}$ in $\mathbb{B}$, $\mathcal{a}_{\mathcal{P}}$ is an action for $(\Sigma_{\mathcal{P}},\Pi_{\mathcal{P}})$. These actions are required to be separately additive in the sense that

$$\mathcal{P},\mathcal{Q} \in \mathbb{B}\ ,\quad \mathcal{P} \wedge \mathcal{Q} = \emptyset \Rightarrow \mathcal{a}_{\mathcal{P}\vee\mathcal{Q}} = \mathcal{a}_{\mathcal{P}} + \mathcal{a}_{\mathcal{Q}}\ , \tag{4.1}$$

and also uniformly Lipschitz continuous with respect to volume, i.e., there exists $k > 0$ such that for every $(P_{\mathcal{B}},\sigma_{\mathcal{B}}) \in \Pi_{\mathcal{B}} \lozenge \Sigma_{\mathcal{B}}$ and every $\mathcal{P} \in \mathbb{B}$,

$$|\mathcal{a}_{\mathcal{P}}(\Psi^{\mathcal{B}}_{\mathcal{P}}P_{\mathcal{B}},\Psi^{\mathcal{B}}_{\mathcal{P}}\sigma_{\mathcal{B}})| \leq k \text{ vol } \mathcal{P}\ . \tag{4.2}$$

In this definition, we interpret the number $\mathcal{a}_{\mathcal{P}}(\Psi^{\mathcal{B}}_{\mathcal{P}}P_{\mathcal{B}},\Psi^{\mathcal{B}}_{\mathcal{P}}\sigma_{\mathcal{B}})$ as the "supply" of $\mathcal{a}$ to $\mathcal{P}$ from the exterior of $\mathcal{P}$ in the process $P_{\mathcal{B}}$ starting at the state $\sigma_{\mathcal{B}}$ of $\mathcal{B}$. It is convenient to drop the subscript $\mathcal{B}$ in all notations and to omit the symbol $\Psi^{\mathcal{B}}_{\mathcal{P}}$ when writing values of such actions. Hence, for example, we shall write (Σ,Π) for $(\Sigma_{\mathcal{B}},\Pi_{\mathcal{B}})$ and $\mathcal{a}_{\mathcal{P}}(P,\sigma)$ for $\mathcal{a}_{\mathcal{P}}(\Psi^{\mathcal{B}}_{\mathcal{P}}P_{\mathcal{B}},\Psi^{\mathcal{B}}_{\mathcal{P}}\sigma_{\mathcal{B}})$. With this notation we can write the equation in (4.1) in the equivalent form

$$\mathcal{a}_{\mathcal{P}\vee\mathcal{Q}}(P,\sigma) = \mathcal{a}_{\mathcal{P}}(P,\sigma) + \mathcal{a}_{\mathcal{Q}}(P,\sigma) \tag{4.3}$$

for all (P,σ) in $\Pi \lozenge \Sigma$. Because $\mathcal{a}_{\mathcal{P}}$ represents an interaction between $\mathcal{P}$ and its exterior, the requirement (4.3) is too strong for applications to bodies in which internal radiation is significant and in which $\mathcal{a}$ represents in some sense "heat added divided by temperature." The term "uniformly" is used in the definition above to express the idea that the number multiplying vol $\mathcal{P}$ in (4.2) does not depend upon the pair (P,σ). This means that the Lipschitz condition with respect to volume cannot be violated by choosing a pathological type of evolution of the body $\mathcal{B}$.

The separate additivity and Lipschitz continuity of the actions in a global action family yield the following result. (See, for example, Theorem 7.2 in the article of BURKILL [1923].

Theorem 4.1. If $(\mathcal{a}_{\mathcal{P}}|\mathcal{P} \in \mathbb{B})$ is a global action family, then for each (P,σ) in $\Pi \lozenge \Sigma$ there exists an integrable function $x \mapsto \mathcal{a}_x(P,\sigma)$ on $\mathcal{B}$ such that

$$\mathcal{a}_{\mathcal{P}}(P,\sigma) = \int_{\mathcal{P}} \mathcal{a}_x(P,\sigma)\,dx$$

for every $\mathcal{P} \in \mathbb{B}$.

The function $x \mapsto \mathcal{a}_x(P,\sigma)$ can be computed at almost every x in $\mathcal{B}$ as a limit of the form

$$\lim_{\substack{\text{vol } \mathcal{P} \to 0 \\ \mathcal{P} \in \mathbb{B}_x}} \frac{\mathcal{a}_{\mathcal{P}}(P,\sigma)}{\text{vol}(\mathcal{P})} .$$

It follows that, at such an x, $\mathcal{a}_x(P,\sigma)$ depends only upon the local process P_x and local state σ_x determined by P and σ. Moreover, relation (2.6) shows that, for every (P'',P') in $\mathbb{P}$ and σ in $\mathcal{D}(P''P')$, there is a subset $\mathcal{S}(P'',P',\sigma)$ of $\mathcal{B}$ such that $\mathcal{B} \setminus \mathcal{S}(P'',P',\sigma)$ has measure zero and the relation

$$\mathcal{a}_x(P''P',\sigma) = \mathcal{a}_x(P',\sigma) + \mathcal{a}_x(P'',P'\sigma)$$

holds throughout $\mathcal{S}(P'',P',\sigma)$. In particular, if a point x lies in every one of the sets $\mathcal{S}(P'',P',\sigma)$, then $\mathcal{a}_x$ is an action for the material point at x.

5. A global statement of the second law and its local consequences.

The algebraic semi-system $(\Sigma_{\mathcal{P}},\Pi_{\mathcal{P}})$ for a part $\mathcal{P}$ of $\mathcal{B}$ may not possess sufficiently many "cycles" $(P_{\mathcal{P}},\sigma_{\mathcal{P}})$ [i.e., pairs in $\Pi_{\mathcal{P}} \lozenge \Sigma_{\mathcal{P}}$ for which $P_{\mathcal{P}}\sigma_{\mathcal{P}} = \sigma_{\mathcal{P}}$] in order to render useful classical statements of the second law which mention only cycles. A version of the second law proposed by COLEMAN & OWEN [1977] to cover such situations suggests the following global statement of the second law.

Second law (global form): There are given a base state σ° of $\mathcal{B}$ and a global action family $(\mathcal{s}_{\mathcal{P}} | \mathcal{P} \in \mathbb{B})$ such that, for every state σ of $\mathcal{B}$ and part $\mathcal{P}$ of $\mathcal{B}$, there exists a number $M(\sigma,\mathcal{P})$ for which

$$P\sigma^\circ = \sigma \Rightarrow \mathcal{s}_{\mathcal{P}}(P,\sigma^\circ) \le M(\sigma,\mathcal{P}) . \tag{(5.1}$$

Condition (5.1) is the assertion that the supply of $\mathcal{s}$ to $\mathcal{P}$ from its exterior in processes taking the body from the base state σ° to any preassigned target state σ is bounded above. The upper bound in (5.1) can depend both on the target state and the part under consideration. For systems with perfect accessibility, a condition of the type (5.1) turns out to be equivalent to the assertion that the supply of $\mathcal{s}$ in cycles is never positive. The latter condition corresponds to the "Clausius Inequality" in classical thermodynamics.

The global form of the second law proposed here cannot be cast directly into local form, because not enough is known about the dependence of $M(\sigma,\mathcal{P})$ upon σ and $\mathcal{P}$. The next result gives a global condition equivalent to the global second law which is more directly suitable for localization.

Theorem 5.1. The global version of the second law is equivalent to the existence of a family of entropy functions, i.e., a family $(S_{\mathcal{P}} : \Sigma_{\mathcal{P}} \to \mathbb{R} | \mathcal{P} \in \mathbb{B})$ such that

$$(P,\sigma) \in \Pi \diamond \Sigma \Rightarrow \mathcal{S}_{\mathcal{P}}(P,\sigma) \leq S_{\mathcal{P}}(P\sigma) - S_{\mathcal{P}}(\sigma) . \quad (5.2)$$

Moreover, if the global version holds, there is a choice $(S^{\circ}_{\mathcal{P}} | \mathcal{P} \in \mathbb{B})$ in which the entropy functions are uniformly Lipschitz continuous with respect to volume:

$$\mathcal{P} \in \mathbb{P},\ \sigma \in \Sigma \Rightarrow |S^{\circ}_{\mathcal{P}}(\sigma)| \leq k \text{ vol } \mathcal{P} , \quad (5.3)$$

and are sub-additive, i.e.

$$\mathcal{P} \wedge \mathcal{Q} = \emptyset \Rightarrow S^{\circ}_{\mathcal{P}\vee\mathcal{Q}} \leq S^{\circ}_{\mathcal{P}} + S^{\circ}_{\mathcal{Q}} . \quad (5.4)$$

We note first of all that, in (5.2), $S_{\mathcal{P}}(P\sigma)$ stands for $S_{\mathcal{P}}(\Psi^{\mathcal{B}}_{\mathcal{P}}(P\sigma))$ and $S_{\mathcal{P}}(\sigma)$ stands for $S_{\mathcal{P}}(\Psi^{\mathcal{B}}_{\mathcal{P}}\sigma)$. Similarly, (5.4) in full detail would read

$$S_{\mathcal{P}\vee\mathcal{Q}}(\Psi^{\mathcal{B}}_{\mathcal{P}\vee\mathcal{Q}}\sigma) \leq S_{\mathcal{P}}(\Psi^{\mathcal{B}}_{\mathcal{P}}\sigma) + S_{\mathcal{Q}}(\Psi^{\mathcal{B}}_{\mathcal{Q}}\sigma)$$

for all σ in $\Sigma_{\mathcal{B}}$. The particular choice $(S^{\circ}_{\mathcal{P}} | \mathcal{P} \in \mathbb{B})$ referred to in (5.3) and (5.4) inherits the Lipschitz volume continuity (with the same Lipschitz constant) from the global action family $(\mathcal{S}_{\mathcal{P}} | \mathcal{P} \in \mathbb{B})$ and inherits from the separate additivity of $(\mathcal{S}_{\mathcal{P}} | \mathcal{P} \in \mathbb{B})$ a property of sub-additivity. This special family of entropy functions is given by the formula

$$S^{\circ}_{\mathcal{P}}(\sigma) = \sup\{\mathcal{S}_{\mathcal{P}}(P,\sigma^{\circ}) | P\sigma^{\circ} = \sigma\} . \quad (5.5)$$

This formula permits one to verify the statements about $(S^{\circ}_{\mathcal{P}}/\mathcal{P} \in \mathbb{B})$ made in the theorem. The relation (5.2) corresponds to the classical "Clausius-Planck inequality."

We are now closer to obtaining a local form of the second law. However, the fact that the functions $(S^{\circ}_{\mathcal{P}} /\mathcal{P} \in \mathbb{B})$ are only sub-additive necessitates an extra step. This is supplied by Theorems 5.1 and 7.2 of BURKILL [1923] which imply that, from the family $(S^{\circ}_{\mathcal{P}}/\mathcal{P} \in \mathbb{B})$ satisfying (5.2)-(5.4), one can produce a family of integrable functions $(x \mapsto s_x(\sigma) \mid \sigma \in \Sigma)$ satisfying the following conditions:

(i) $(\sigma \mapsto \int_{\mathcal{P}} s_x(\sigma)dx \mid \mathcal{P} \in \mathbb{B})$ is a family of entropy functions;

(ii) $S^{\circ}_{\mathcal{P}}(\sigma) \leq \int_{\mathcal{P}} s_x(\sigma)dx$ for all $\sigma \in \Sigma$.

The important point here is that the family in (i) is not only uniformly Lipschitz continuous with respect to volume, but is separately additive. In other words, the family inherits both the properties of the global action family $(\mathcal{A}_{\mathcal{P}}|\mathcal{P} \in \mathbb{B})$. Condition (i) may now be used directly to establish the following theorem.

<u>Theorem 5.2</u>: Let σ° and $(\mathcal{A}_{\mathcal{P}}|\mathcal{P} \in \mathbb{B})$ satisfy the global second law. As in Theorem 4.1, for each (P,α) in $\Pi \diamond \Sigma$, let $x \mapsto \mathcal{A}_x(P,\sigma)$ denote an integrable function satisfying

$$\mathcal{A}_{\mathcal{P}}(P,\sigma) = \int_{\mathcal{P}} \mathcal{A}_x(P,\sigma)dx \tag{5.6}$$

for every $\mathcal{P}$ in $\mathbb{B}$. For each state σ of $\mathcal{B}$ there is a subset $\mathcal{S}(\sigma)$ of the interior of $\mathcal{B}$ with the same measure as $\mathcal{B}$ and, for each x in $\mathcal{S}(\sigma)$, there is a number $M_x(\sigma)$ such that, whenever $P\sigma^{\circ} = \sigma$, the relation

$$\mathcal{A}_x(P,\sigma^{\circ}) \leq M_x(\sigma) \tag{5.7}$$

holds for almost every x in $\mathcal{S}(\sigma)$. Moreover, $\mathcal{A}_x(P,\sigma)$ and $M_x(\sigma)$ depend only upon the local quantities (P_x,σ_x) and σ_x , respectively, and, for each state σ of $\mathcal{B}$, $x \mapsto M_x(\sigma)$ is integrable

We call (5.7), together with the integrability of the functions $x \mapsto M_x(\sigma)$, the <u>local form of the second law</u>, and Theorem 5.2 tells us that the global form of the second law implies the local form. It is clear that the local form implies the global form, so we now have the main result of this paper.

Theorem 5.3: Let σ° be a base state of (Σ,Π), and let $(\mathcal{U}_\mathcal{P} | \mathcal{P} \in \mathbb{B})$ be a global action family. It follows that the global and local forms of the second law are equivalent.

We note that the condition (5.2) has a corresponding local form

$$(P,\sigma) \in \Pi \diamond \Sigma \Rightarrow \mathcal{s}_x(P,\sigma) \leq s_x(P\sigma) - s_x(\sigma) \quad , \qquad (5.8)$$

with the functions $x \mapsto \mathcal{s}_x(P,\sigma)$ obeying (5.6); the inequality in (5.8) holds throughout $\mathcal{B}$ with the exception of a set of measure zero which may depend upon both P and σ. In fact, (5.8) follows from condition (i) and implies that (5.7) holds with

$$M_x(\sigma) = s_x(\sigma) - s_x(\sigma^\circ) \quad .$$

Conversely, (5.8) implies (5.2) with

$$S_\mathcal{P}(\sigma) = \int_\mathcal{P} s_x(\sigma)\,dx \quad .$$

Therefore, the local form of the second law (5.7), the global form of the second law (5.1), the global "Clausius-Planck inequality" (5.2), and the local "Clausius-Planck inequality" (5.8) are equivalent statements.

I wish to acknowledge the encouragement and interest of Professor F. W. Lawvere as well as the support of the National Science Foundation during the preparation of this paper.

References

1923 Burkill, J. C., Functions of intervals, Proc. London Math. Soc. (2), **22**, 275-310.

1959 Noll, W., The foundations of classical mechanics in light of recent advances in continuum mechanics, pp. 261-281 of The Axiomatic Method, with Special Reference to Geometry and Physics (Symposium at Berkeley, 1957), Amsterdam, North-Holland Publishing Co.

1963 Noll, W., La mecanique classique, basée sur un axiome d'objectivité, pp. 47-56 of La Méthode Axiomatique dans les Mécaniques Classiques et Nouvelles (Colloque International, Paris, 1959), Paris, Gauthier-Villars.

1967 Gurtin, M. E., and W. O. Williams, An axiomatic foundation for continuum thermodynamics, Arch. Rational Mech. Anal. **26**, 83-117.

1969 Day, W. A., A theory of thermodynamics for materials with memory, Arch. Rational Mech. Anal. **34**, 85-96.

1970 Williams, W. O., Thermodynamics of rigid continua, Arch. Rational Mech. Anal. **36**, 270-284.

1972 Noll, W., A new mathematical theory of simple materials, Arch. Rational Mech. Anal. **48**, 1-50.

1972 Wilmanski, K., On thermodynamics and functions of states of nonisolated systems, Arch. Rational Mech. Anal. **45**, 251-281.

1974 Coleman, B. D., and D. R. Owen, A mathematical foundation for thermodynamics, Arch. Rational Mech. Anal. **54**, 1-104.

1974 Williams, W. O., unpublished invited lecture at the meeting of the Society for Natural Philosophy, Pisa.

1977 Coleman, B. D., and D. R. Owen, On the thermodynamics of semi-systems with restrictions on the accessibility of states, Arch. Rational Mech. Anal. **66**, 173-181.

THERMODYNAMICS AND THE HAHN-BANACH THEOREM

by

Martin Feinberg
Department of Chemical Engineering
University of Rochester

Richard Lavine
Department of Mathematics
University of Rochester

In traditional thermodynamics, the second law is usually stated so as to apply to every conceivable process, but its implications are studied carefully only for very simple processes, such as reversible homogeneous ones. One of the principal conclusions is the definition, for each state which points of the body pass through during such processes, of two quantities, the <u>thermodynamic temperature</u> θ , and the <u>specific entropy</u> η , which satisfy a relation called the <u>Clausius-Duhem inequality</u>. The total entropy of a homogeneous body is the product of its mass times the specific entropy η . The Clausius-Duhem inequality says roughly that the integral over the process of the quantity of heat gained, divided by the temperature of the points which receive the heat, is no greater than the change in total entropy.

In continuum mechanics, inhomogeneous irreversible processes are studied, and entropy, temperature, and the Clausius-Duhem inequality play an important role. Thus it is natural to ask whether these can be derived from a statement of the second law in a context general enough to include continuum mechanics. We sketch here a way of doing this. A more detailed summary is given in [1], and a full treatment of temperature appears in [2].

In a continuum theory a material point is completely described by giving its <u>state</u>, typically a finite collection of quantities like pressure and density. We take the set of possible states to be a compact Hausdorff space Σ . Thus a process would be specified by a function mapping the product of the set of material points with a time interval $[t_0,t_1]$ into Σ . But for our purposes we only need to keep track of the total mass of material in each state at the beginning of the process (time t_0) and at the end (time t_1) . These are given by positive Borel measures m_0 and m_1 on Σ ; for $B \subset \Sigma$, $m_j(B)$ is the total mass occupying states in B at time t_j .

Thermodynamics enters when we consider heat transfer between the body and its environment. Again we need not consider the details of where and when this transfer happens, but we do need to keep track of the <u>states</u> in which heat is received by parts of the body. This information is given by the <u>heating measure</u> q , a signed Borel measure on Σ . For $B \subset \Sigma$, $q(B)$ represents the net heat gained during the process by material points in states contained in B .

In fact all the necessary information about a process is given by the pair of signed measures $(m_1 - m_0, q)$ on Σ . Such pairs we regard as elements of the locally convex topological vector space $M(\Sigma) \oplus M(\Sigma)$, where $M(\Sigma)$ is the set of signed Borel measures on Σ with the weak-star topology. In traditional thermodynamics one considers "ideal processes" which are impossible, but can be approached by actual processes. The counterparts to these are the (weak-star) limits of pairs of measures $(m_1 - m_0, q)$ corresponding to actual processes. It is possible to argue on physical grounds that the set of all such measures forms a non-empty closed convex cone P in $M(\Sigma) \oplus M(\Sigma)$.

There are many statements, not all equivalent, which are called the "second law of thermodynamics." The Kelvin-Planck version says roughly that it is impossible for a body undergoing a cyclic (possibly ideal) process to convert heat energy completely into work. The first law equates the work performed by a body in a cyclic process with the net heat it gains from the environment, so the Kelvin-Planck law says that if some heat is gained in a cyclic process, some must also be lost. This can be given a precise meaning in terms of the pairs of measures introduced above:

1) A process is cyclic if $m = m_1 - m_0 = 0$. (At the end of the process there is the same amount of material in each state as at the beginning.)
2) Some heat is gained and none lost if q is a positive, non-zero measure.

Thus we say that P satisfies the Kelvin-Planck condition if P contains no pair $(0,\nu)$ in which ν is strictly positive. This implies that the closed convex cone P is disjoint from the compact convex set $K = \{(0,\nu): \nu > 0, \nu(\Sigma) = 1\}$. Then the Hahn-Banach theorem provides a separating linear functional, which is positive on K and non-positive on P . This linear functional $f: M(\Sigma) \oplus M(\Sigma) \to \mathbb{R}$ is given by a pair of continuous functions η and β on Σ: $f(\mu,\nu) = \int \beta d\nu - \int \eta d\mu$. It is easy to show that β must be positive. We interpret $1/\beta = \theta$ as temperature and η as entropy, and the separation inequality becomes

$$\int \frac{dq}{\theta} \leq \int \eta d(\Delta m)$$

which is a precise form of the Clausius-Duhem inequality.

Thus the existence of entropy and temperature functions meaningful for continuum mechanics becomes a consequence of very general assumptions and basic functional analysis.

Of course our η and θ need not be unique. In fact we show that they are essentially unique if and only if P contains sufficiently many elements which correspond to reversible processes. The fact that uniqueness implies a large supply of reversible processes uses the Hahn-Banach theorem. And the same technique can be used to relate various properties of the functions η and θ to the types of elements contained in P . In particular, in [1,2] we consider what it means for one state to be "hotter than" another in terms of the processes represented in P , and we examine the extent to which this partial ordering is reflected by the values of the temperature functions θ .

REFERENCES

1. Feinberg, M. and Lavine, R., Foundations of the Clausius-Duhem inequality in C. Truesdell, Rational Thermodynamics, 2nd edition, to appear.

2. Feinberg, M. and Lavine, R., Thermodynamics based on the Hahn-Banach Theorem: the Clausius inequality, to appear, Arch. Rat. Mech., 1983.

WHAT IS THE LENGTH OF A POTATO?

An Introduction to Geometric Measure Theory

Stephen H. Schanuel

Department of Mathematics
State University of New York at Buffalo
Buffalo, New York 14214

The question in the title probably sounds a bit peculiar; but I hope to persuade you that it has a unique sensible interpretation, and to show you several ways (at least for a potato shaped like a ball) to compute the answer. But my real goal is more ambitious: I hope to reform your intuition about geometry, to get you to incorporate into your picture of Euclidean geometry the sweeping changes in fundamental notions stemming from the work of Euler, Gauss, Riemann, Minkowski, and many others. For this reason I say very little about proofs (except to indicate where they can be found), and try to show the ideas in the simplest setting where they make their appearance.

Our topic is volume, area, length, and number. We begin with length. Imagine an idealized measuring stick, say of length one inch as pictured below. (I have drawn the heavy dots to emphasize that I'm thinking of a closed segment.)

Now this is really a rather poor instrument for measuring lengths. The defect is that if we magnify the segment by a factor of two, the resulting segment is not the disjoint union of two copies of the original; the two pieces have a one-point overlap.

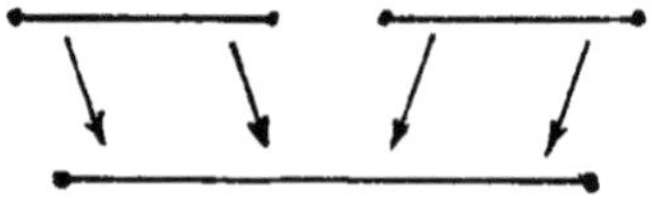

This suggests that our original segment was infinitesimally larger than one inch; its true size is 1 in + 1, the 1 for the one extra point The basic lesson to be drawn from the geometers since Euclid is that it is not only possible, but even desirable, to keep track of this infinitesimal excess. So the "total size" of a solid figure in Euclidean space should not be a pure volume, but a formal sum of terms volume + area + length + number (so formally polynomials in the quantity in=inch) Let's calculate some examples:

1) A closed line segment of length L inches has size Lin + 1.

2) A closed rectangle has size

Win + 1

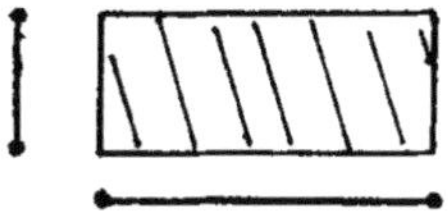

Lin + 1

$$(\mathrm{Lin} + 1)\ (\mathrm{Win} + 1) = \mathrm{LWin}^2 + (\mathrm{L} + \mathrm{W})\mathrm{in} + 1$$

3) A right triangle has its size computed as Euclid did, except that to take account of the excess

Bin + 1

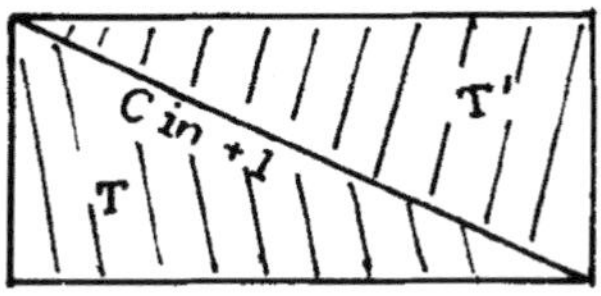

Ain + 1

we must use

$$\mathrm{Size}(T \cup T') = \mathrm{size}T + \mathrm{size}T' - \mathrm{size}(T \cap T')$$

$$(\mathrm{Ain} + 1)(\mathrm{Bin} + 1) = 2\mathrm{size}T - (\mathrm{Cin} + 1)$$

$$\mathrm{size}T = \frac{AB}{2}\mathrm{in}^2 + \frac{A+B+C}{2}\mathrm{in} + 1$$

By now, perhaps you have begun to guess the significance of the terms in the size. The "area" is just the area as Euclid would have computed it. The "length" is one-half the perimeter. (One explanation for the factor ½: the boundary is only half exposed, so that for a two-dimensional creature to paint the exposed boundary requires only half as much paint as if he were to paint the one-dimensional figures which are the boundaries of our rectangle and triangle. These boundaries, as geometric figures in their own right, have their usual lengths.) The "number" of the figure is what came to be called the "Euler characteristic" after Euler's proofs that the number of a two-sphere is 2, and some investigations of one-dimensional figures.

Before going further, we must look a bit more closely at the number of a figure. Since the time of Euclid, there have been two great advances in our notion of cardinal number. From Cantor we learned to count infinite discrete sets, and from Euler we learned to count extended bodies. Of these two advances, Euler's has been by far the more important; but we seem, most of us, to have spent more effort retraining our intuitions to incorporate Cantor's ideas than Euler's.

Let's try to remedy that, at least a little, now. First an elementary observation about counting:

number$(A \cup B)$ = number (A) + number (B) - number $(A \cap B)$

as this pile of potatoes illustrates

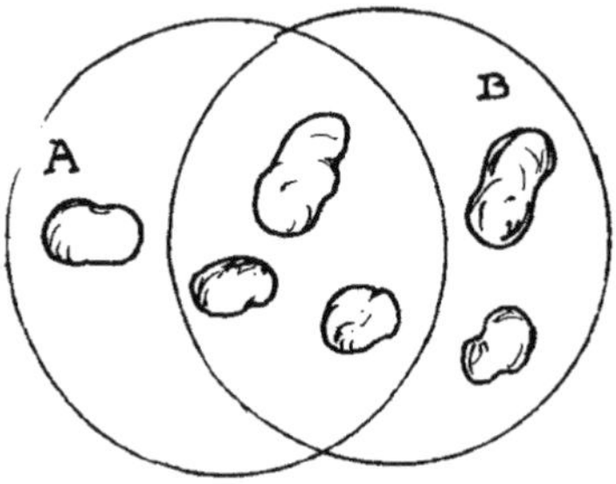

$6 = 4 + 5 - 3$

Of course any child could observe that, but how many of them have observed that the next example illustrates the same phenomenon?

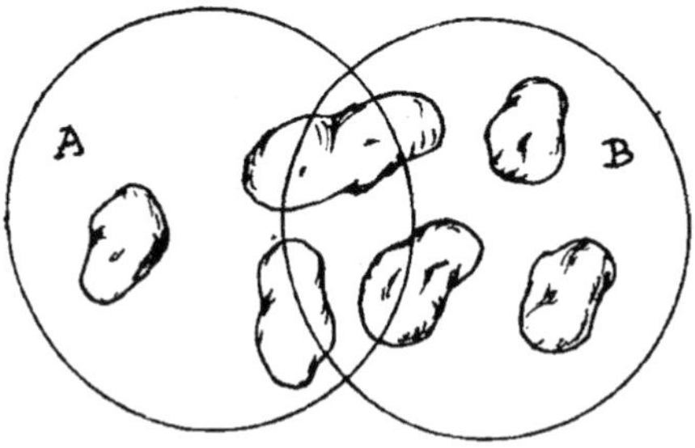

$6 = 4 + 5 - 3$

Each object, be it a small potato or a large one or even a piece of a potato, counts as one.

We seem to get into trouble if our pile includes a doughnut:

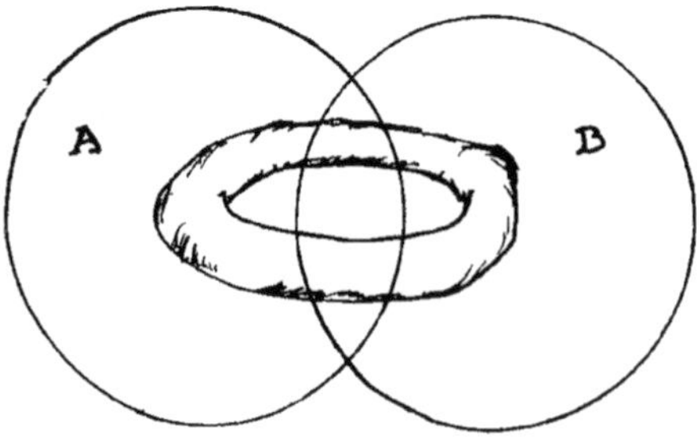

number $(A \cup B)$ = number (A) + number (B) - number $(A \cap B)$

$= 1 + 1 - 2 = 0$

So we're forced to count a doughnut as zero, if we want counting to be finitely additive when an extended body (or pile) is written as a union of parts which are not clopen. Of course, we're neglecting, for now, the important question of what sorts of bodies and what sorts of parts are to be allowed; what is apparent is that some sort of "combinatorial finiteness" is needed. To avoid these difficulties, let's restrict our attention for the moment to finite compact polyhedra, for which there is no difficulty in making precise the definitions of number, etc. and in proving the basic propositions. (But we reserve the right to draw examples from more general cases which have been worked out over the past century or so.)

We should illustrate at least one use of this refined notion of size, Steiner's formula. Even supposing one is interested only in the area of plane figures, one can ask for the area of the set of all points at distance at most R inches from a convex plane region T .

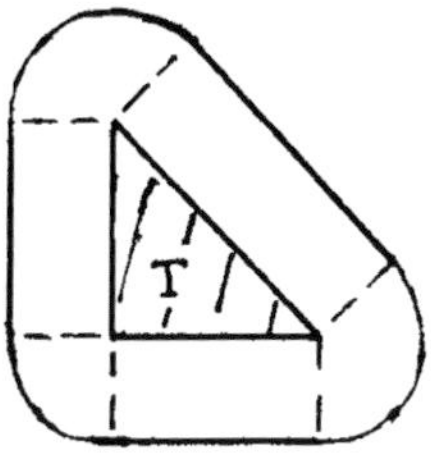

In the picture, it's clear that the large region decomposes as $T \cup (\text{rectangles}) \cup (\text{sectors of a disc})$, and the total area is

Total area = $1 \cdot \text{Area}(T) + (2R\ \text{in})\ \text{Length}(T) + (\pi R^2 \text{in}^2)\ \text{Number}(T)$,

recalling that the length of T is one-half its perimeter. This is quite general, for any compact convex set in N dimensions, and is Steiner's formula. (The coefficients are just the n-dimensional measure of a ball of radius R in in n-space, here for n = 0, 1, 2.) The right side of Steiner's formula computes something even if T is not convex: one must think of the left side as the N-dimensional volume-integral of the function whose value at any point p is the number (=Euler characteristic) of the intersection of T with the closed ball of radius R in centered at p . Of course, when T is compact convex, this intersection is also, so this number is either 1 or 0, and the function becomes the characteristic function of the large region. This illustrates a general phenomenon in the whole subject: all problems reduce to the problem of correctly understanding number; length, area, etc. then are relatively easy to understand.

Another illustration of the primacy of number comes from the <u>integral-geometric interpretation</u> of length, etc. for, say, a figure in 3-space. To calculate the length (= 1-dimensional measure) look at all 2-planes in 3-space (because 2 = 3 - 1). Now on the space of planes there's a measure, unique up to a constant factor, invariant under rigid notions, giving open sets of planes positive measure and compact sets of planes finite measure; normalize it so that the set of planes meeting a line segment of length 1 inch has measure 1 inch. (It is best to think of this measure as valued in lengths, rather than pure numbers.) Now to calculate the length of our figure, simply integrate, over the space of planes, the function whose value at any plane is the "number of times the plane hits the figure", which must of course be interpreted as the Euler characteristic of the intersection of the plane with the figure. Similarly, to calculate the area of a figure, normalize the (area-valued) measure on the space of lines so that the set of lines meeting a square of side 1 in has measure 1 in^2, and proceed as before.

Returning now to the example which illustrated Steiner's formula, we can notice a bit more. The area, length, and number are not merely quantities associated to our triangle, but are the integrals, or total measures, of measures supported on the triangle: the area measure is the usual one; the length measure is one-half the usual length measure on the edges; and the number measure associates to the lower left vertex the measure ¼ and to each of the other vertices the measure 3/8, corresponding to the fractions of disc situated at each vertex in our picture. Federer has shown that for a class of subsets of Euclidean space called "sets of positive reach", significantly generalizing closed convex sets, one can use this idea to precisely define the measures, and to prove their relevant properties. For closed convex sets, Minkowski studied these quantities, which he called "Quermassintegralen". Unfortunately, he normalized them and indexed them in such a way as to obscure their geometric interpretation as length, area, etc.

The formula for the zero-dimensional measure is called the "Gauss-Bonnet formula", especially in the case of smooth manifolds with boundary. One example, simpler than our

number(isosceles right triangle) = ¼ + 3/8 + 3/8 = 1

is familiar to all children. To count the number of pieces of rope in a tangled mess of rope, it is unnecessary to separate the pieces; the number of the pile is concentrated at the ends of the pieces, each end counting one-half. It is a short step, conceptually, from this to the idea that the number of a solid object is also the integral of some local quantity; for a 3-manifold with boundary, for example a potato, in Euclidean 3-space, the Euler characteristic is the integral of a measure dm_o concentrated on the surface of the body:

$$dm_o = (4\pi)^{-1} R_1^{-1} R_2^{-1}\, ds,$$

where ds is the usual surface area measure, and R_1 and R_2 are the "principal radii of curvature" at a point. Note that since R_1 and R_2 have the dimensions of length, our measure has dimension $(\text{length})^{-2}\cdot$ area, so is a pure number, as it should be. This formula generalizes to give formulas for the k-dimensional measures dm_k for an n-manifold M with smooth boundary ∂M in n-space, for $k = 0,\ldots,n-1$;

$$dm_k = c_{n,k}\, p_{n-1-k}(R_1^{-1},\ldots,R_{n-1}^{-1})\, ds$$

where p_i is the i-th elementary symmetric function (homogeneous of degree i), ds is the usual (n-1)-dimensional surface area measure, and $c_{n,k}$ is a constant which can easily be computed, for example, by specializing to the case of a ball, for which we know m_k from Steiner's formula. For instance, for the potato, or solid body in 3-space,

$$dm_2 = \tfrac{1}{2}\, ds,$$
$$dm_1 = (2\pi)^{-1}(R_1^{-1} + R_2^{-1})\, ds,$$

dm_0 was calculated above, and dm_3 is the usual volume measure restricted to M . (It is a peculiarity of smooth figures that the lower-dimensional measures are spread all over the boundary. For polyhedra, dm_k is concentrated on the k-cells; but if you imagine approximating M by polyhedra you see why the measures get spread out.)

The observation that $m_0(S^n)$, the Euler characteristic of the n-sphere, is 2 if n is even, 0 if n is odd, generalizes.

$$dm_k(\partial M) = \begin{cases} 2\, dm_k(M) & \text{if } n-k \text{ is odd} \\ 0 & \text{if } n-k \text{ is even.} \end{cases}$$

Hence if we take n odd and $\partial M = \emptyset$, so a manifold without boundary, then not only is $\int dm_0 = m_0(M) = 0$, but in fact $dm_0(M)$ is <u>identically</u> zero, so $\int f dm_0(M) = 0$ for any integrable function f . More generally, for a manifold M without boundary, $dm_k(M) = 0$ in all odd codimensions, since it's $\tfrac{1}{2}\, dm_k(\partial M)$. Thus for instance for a 2-manifold with boundary (say in 3-space) the length measure is just one half the length measure on the boundary curves, and not spread over the surface; while the number measure is spread all over, like the area measure.

Let's look at one more example, to help visualize the measures: a solid cylinder M of radius R and height H . (This is topologically, though not smoothly, a manifold with boundary; so all of the preceding paragraph applies to it). $M = D\times I$, where D is a disk of radius R, and I an interval of length H . So the total measure is given by

$$m(M) = (m_2(D) + m_1(D) + m_o(D))(m_1(I) + m_o(I))$$
$$= (\pi R^2 + \pi R + 1)(H + 1)$$
$$= \pi R^2 H + (\pi R^2 + \pi RH) + (\pi R + H) + 1$$

where the homogeneous term of degree k gives $m_k(M)$. The formulas for m_3, m_2, and m_o are just the usual ones for volume, half of surface area, and Euler characteristic; but note the convenience of combining the terms into a single "polynomial", even for computing the surface area. More interesting is the analogous formula for dm ; for instance the length measure

$$dm_1(M) = dm_o(D) \times dm_1(I) + dm_1(D) \times dm_o(I).$$

Thus the length measure on a cylinder is the sum of two simpler (product) measures:

$dm_o(D) \times dm_1(I)$ is the product of the measure uniformly distributed over the circle ∂D, with total measure (the pure number) 1, multiplied by the length measure on the interval; so this term is uniformly distributed over the lateral surface of our cylinder, with total measure H (a length).

$dm_1(D) \times dm_o(I)$ is the product of the length measure on D, which is uniformly distributed over the circle ∂D, giving each arc a measure of ½ its length, multiplied by the pure number measure on I which gives each endpoint measure ½ ; so this term is concentrated on the top and bottom rims of our cylinder, and gives to each arc measure ¼ its length.

Notice that if we fix H and let R tend to zero, so that our cylinder tends to a line segment of length H, then the measures $dm_k(M)$ tend to those for the segment. This is an instance of a general continuity property of the measures, but the correct formulation of the appropriate notion of convergence of variable figures has been worked out only in special cases, for example for sets of positive reach by Federer. For compact convex sets, things are particularly simple, as Minkowski already knew: the sets are close if and only if they're close in the Hausdorff metric

$$d(A,B) = \sup(\{d(a,B),\ a \in A\} \cup \{d(b,A),\ b \in B\}).$$

Indeed, for (compact) convex sets A,B, the measures have many nice properties, for instance: $dm_k(A) \geq 0$; $A \subset B$ implies $m_k(A) \leq m_k(B)$.

Thus the length of B is greater than the length of A ; clearly this needn't hold if A and B are not convex, as a long spring in a small cylinder demonstrates; and $dm_k(A) \geq 0$ is false even for k = 0 for non-convex A . For non-convex A , there's a version of positivity: $m(A) \geq 0$, not coefficientwise as for convex A , but only in the ordering in which the term of highest degree dominates. For convex A , a lot is known about the possible values of the vector $(1=m_o(A), m_1(A), \ldots, m_n(A))$; the isoperimetric inequality is one constraint, and others are known, but I do not believe that a complete description of the image of this vector as A ranges over compact convex sets is known, even for n = 3 .

To realize the utility of having the measures dm_k , instead of just the total $m_k = \int dm_k$, consider Pappus' formulas for the volume and surface area of a solid of revolution. These, you will recall, say that if we revolve the plane figure K (in the upper half-plane) about the x-axis to give a solid $\tilde{K}$, then

$$m_3(\tilde{K}) = m_2(K)\cdot 2\pi\, y_2 \quad \text{and}$$

$$m_2(\tilde{K}) = m_1(K)\cdot 2\pi\, y_1 ,$$

where y_k is the result of averaging the distance y of a point from the x-axis, with respect to the measure $dm_k(K)$. (Of course, the second formula is usually multiplied by 2 , then saying that the surface area $2m_2(\tilde{K})$ is the perimeter $2m_1(K)$ times the average over the boundary curve of the y-coordinate times 2π .) But the important fact to notice is that y_k cannot be computed from m_k ; we need to know dm_k to do the averaging. Putting Pappus' theorems in this form immediately suggests another theorem:

$$m_1(\tilde{K}) = m_o(K)\cdot 2\pi\, y_o$$

This is also true, unless K meets the axis of revolution in a set of positive length $L = m_1(K\cap \text{axis})$; then L must be added to the right side. For one dimension lower, the main term disappears, but the correction does not:

$$m_o(\tilde{K}) = m_o(K\cap \text{axis}),$$

so the general form is

$$m_k(\tilde{K}) = m_{k-1}(K)\cdot 2\pi\, y_{k-1} + m_k(K\cap \text{axis}) .$$

It's now easy to use these formulas to get alternative calculations of m_k for ball and cylinder, as well as for cones and so on.

All this has been greatly generalized, but much remains to be done. We have not emphasized the fact that a figure K has its own intrinsic (geodesic) metric $d_K(x,y)$, in terms of which the measures dm_k should have an invariant description, which I don't know how to give in a nice way. Closely connected with this are two other problems: what characterizes the metric spaces K which bear these measures, and how does one describe closeness between such K's?

I hope that our parade of familiar objects viewed in terms of their associated measures dm will have persuaded you that the "length of a potato" is a useful notion, and that these lower dimensional terms in the measure of a solid are simple enough to be taught in elementary calculus. I have performed the experiment; some of my students enjoyed it.

I have left one mystery to the end: what is the actual value of the length of a ball? You can work it out by calculating the volume of a ball of radius R + S by applying Steiner's formula to a ball of radius S . Or you can use our formulas for solids of revolution. Or you can use the integral-geometric approach. It's twice the diameter.

BIBLIOGRAPHY

There is a vast literature on this topic, under the headings of differential geometry, Riemannian geometry, convexity, geometric measure integral geometry, and more. As an outsider to all of these, I found three sources most helpful. One is conversation with Bill Lawvere; indeed my own investigations began when he and I asked ourselves: in what sense is the boundary of a solid the derivative of a solid? (I should add that he does not profess to be an expert in the areas listed, any more than I do.) The other two most helpful sources are listed below; Federer's paper gives complete proofs of most of what we have asserted, for the rather general case of "sets of positive reach in Euclidean space", while Hadwiger's book is a leisurely elementary account of the basic notions, especially for polyhedra.

Federer, H. Curvature Measures, Trans. Amer. Math. Soc. 93 (1959) 418 - 491.

Hadwiger, H. Vorlesungen über Inhalt, Oberfläche, und Isoperimetrie, Springer-Verlag, Heidelberg, 1957.